THE MASTER HANDBOOK OF ALL HOME HEATING SYSTEMS—
Tuneup, Repair, Installation & Maintenance

Other TAB books by the authors:

No. 820 *Central Heating and Air Conditioning Repair Guide*
No. 904 *Homeowner's Guide to Saving Energy*

THE MASTER HANDBOOK OF ALL HOME HEATING SYSTEMS—
Tuneup, Repair, Installation & Maintenance

BY BILLY L. PRICE & JAMES T. PRICE

TAB BOOKS Inc.
BLUE RIDGE SUMMIT, PA. 17214

FIRST EDITION

FIRST PRINTING—SEPTEMBER 1979
SECOND PRINTING—MARCH 1981

Copyright © 1979 by TAB BOOKS Inc.

Printed in the United States of America

Reproduction or publication of the content in any manner, without express permission of the publisher, is prohibited. No liability is assumed with respect to the use of the information herein.

Library of Congress Cataloging in Publication Data

Price, Billy L.
    The master handbook of all home heating systems.

    "TAB" # 1176."
    Includes index.
    . Heating—Equipment and supplies—Maintenance and repair—Amateurs' manuals. I. Price, James Tucker, 1955-joint author. II. Title.
TH7225.P7   697'.0028   79-17177
ISBN 0-8306-9757-8
ISBN 0-8306-1176-2 pbk.

Preface

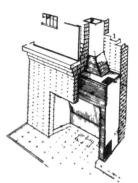

Each winter we are painfully reminded that heating costs are rising higher and higher. This book was written to help you do something to cut your rising fuel bill.

For many homeowners—even those who are good do-it-yourselfers—the heating system seems like an immensely complicated device and one that they shouldn't attempt to work on. But this is a wrong assumption. This book will acquaint you with all types of heating systems. Step-by-step diagrams and plain, simple language lead you through the operation of the different types of heating systems and show you the basic maintenance procedures you can use to keep your heating system running at peak efficiency.

If you are a person considering doing heating system planning, tuneups and servicing on a part-time basis, we think this book provides an excellent introduction to heating systems and heating system maintenance. You will find heating system maintenance to be a skill much in demand, especially as more and more fuel-conscious homeowners request annual heating system tuneups.

There are a number of ways to save energy and money on a home heating system. This book outlines all of them for you in a straightforward manner. It will show you not only how to select the best system or combination of systems to save the most on your heating bill, but it will show you step-by-step how to install it and how to keep it tuned up. You may be surprised to learn that many so-called "energy saving" ideas will actually cost you more than they save. This book tells you about many of them.

The purpose of this book is to save you money each day of the heating season. You will find chapters on planning for a new heating system, modifying your present system to lower your heating costs, adding room heaters or wood units, and tuning up and servicing your heating system to make it operate much more efficiently. Throughout the book you will find hundreds of ways you can save money, and virtually every one can be easily performed by the do-it-yourself homeowner.

That's why we like to think this is a book no homeowner should be without.

Billy L. Price
James T. Price

Contents

1 Basic Tools, Testers and Troubleshooting..................................11
Voltmeter—Ohmmeters—Closed, Short and Open Circuits—Ammeters

2 What You Should Know About Your Heating System...............31
Air Flow Through the Furnace—Upflow, Downflow and Horizontal Furnaces—Heating-Cooling Furnaces—Thermostats—Safety Controls—Furnace Operation—Hydronic Heating Basics

3 Selecting the Lowest Cost Fuel 63
Square One: Know Your Questions—Figuring the Fuel Costs—Fuel Efficiency and Heat Pumps

4 General Furnace Tuneup Tips..73
Filters—Cleaning the Furnace Fan—Adjusting the Fan Control—More on Furnace Tuneups—Energy Efficiency Checklist

5 Tuneup and Servicing Techniques for Oil, Gas, and Electric Furnaces and Boilers................... 109
Oil Furnaces—Oil Furnace Operation—Fuel Tank and Filter—Oil Burner Tuneups—Oil Burner Tuneup Checklist—Further Oil Burner Servicing and Troubleshooting—Gas Burners—Gas Burner Operation—Pilot Lights, Thermocouples and Electric Ignition—Gas Burner Tuneup—Gas Burner Tuneup Checklist—Troubleshooting and Servicing Gas Burners—Electric Furnaces—Servicing the Electric Furnace

6 Furnace Fuel Conversions to Cut Heating Costs 155
Coal Furnaces—Adding Wood and Coal Heaters to Your Heating System—Gas Furnaces—Oil Furnaces—When Installing a New Furnace—Adding Heat Pumps

7 Supplementary Heaters 173
Types of Supplementary Heaters—Installing Electric Room Heaters—Thermostats—Baseboard Heater Troubleshooting and Tuneup—Summary

8 The Duct System: The Final Link 183
Duct System Basics—Duct Tools—Simple Ductwork and Energy Savings—Duct Registers and Their Placement—Adjusting the Registers—Insulating the Duct System—Summary

9 Saving Money With Wood and Coal Heat 201
Can You Save Money?—Wood and Coal Stoves—Types of Stoves—Buying a Stove—Installing a Stove—Other Stove Installation Ideas—Attaching a Stove to Your Central Heating System—Stove Maintenance—Wood and Coal Furnaces—Fireplaces—Parts of a Fireplace—Fireplace Safety—Prefabricated Fireplaces—Increasing Fireplace Heat Output—Checklist

10 Heat Pumps 243
Increasing Popularity—Will a Heat Pump Save You Money?—Heat Pump Operation—Supplementary Heat and the Balance Point—Other Types of Heat Pumps—Heat Pump Efficiency—More Buying Tips—Maintenance

11 Thermostats 267
Types of Thermostats—Outdoor Thermostats—Thermostat Maintenance—Thermostat Anticipator—Thermostat Location—Thermostat Set-back—Guidelines

12 Humidifiers 291
Humidity Basics—Humidifiers and Fuel Savings—Types of Humidifiers—Selection and Sizing—Installation—Wiring the Humidifier and Humidistat—Humidifier Maintenance

13 Homeowner's Energy Tax Credit and Other Conservation Incentives 321
How the Credit Works—Some Basic Rules—Items That Qualify—State and Local Government Energy Incentives

Glossary 327

Appendix A 339

Appendix B 341

Index 347

Basic Tools, Testers and Troubleshooting

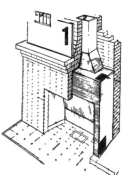

One of the first things you'll learn about your heating system is that it doesn't require a lot of fancy tools to do most of the routine servicing yourself. That means double savings for you. You save money on heating and servicing bills by doing routine maintenance and servicing yourself. And, you can also keep those savings and not have to invest them in a lot of fancy servicing equipment that will take 5 years to pay back.

Of course, heating system servicing is just like any other type of mechanical servicing—the more complicated the task, the more you'll need specialized tools and equipment to do it right. But most heating system routine servicing can be done with a minimum of special tools, and for the more complicated jobs requiring special tools, you'll probably want to call the serviceman anyway.

If you're a handy do-it-yourselfer, you probably already have an assortment of basic tools that you'll find necessary for work on your heating system. Screwdrivers, wrenches, wire cutters, pliers and electrical tape are necessary for servicing a heating system. You already know how to use those tools. You will find tin snips, an electric drill, a pop-rivet gun and furnace duct tape necessities if you do much work on your duct system. And for basic tuneup and servicing, you should have a voltmeter, an ohmmeter, and perhaps an ammeter and pressure gauge.

If you're unfamiliar with your heating system and have not done any work with it before, you're probably hesitant to take the plunge and purchase *any* new tools until you have a chance to see how much you will use them. That's fine, and to make taking the plunge easier, the two most basic tools—the voltmeter and the ohmmeter—can be replaced with inexpensive homemade testers that let you get your feet wet in servicing your heating system without laying out more than a few dollars cash!

VOLTMETER

The voltmeter is used to check the amount of electrical power reaching a component. By placing the voltmeter leads on the power terminals of the component, you can determine whether power is reaching it. If you suspect, for instance, that your furnace fan is not operating properly, you can tell whether there is power in the wires coming to the fan by shorting across the fan switch and touching the voltmeter probes to the fan motor terminals. The voltmeter measures the power reaching the component, so the unit must be plugged in and any power controls (the fan switch in the example above) must be closed so that the unit is "on". Once you know whether voltage is reaching the component, you know whether the problem is with the component and its control or with the system that sends voltage to it. There are several different types of voltmeters available, ranging from a simple neon bulb that glows when there is voltage across two terminals (Fig. 1-1) to a more expensive meter with several scales that will tell you exactly how many volts are in the circuit (Fig. 1-2). The latter type of unit is usually combined with an ohmmeter (discussed later), and the cost of such a meter can run between $10 and $100, depending on the model and the number of accessories. A neon bulb costs about $5 or less, but this type of voltage indicator will not tell you whether 90 volts or 120 volts are reaching the terminals. Such voltage figures can be crucial when checking an electric motor. Therefore, if you are checking voltage in a situation where an accurate voltage reading is needed, you will need a meter.

A voltmeter has a dial that sets the meter to read a number of different scales. If the proper voltage across two

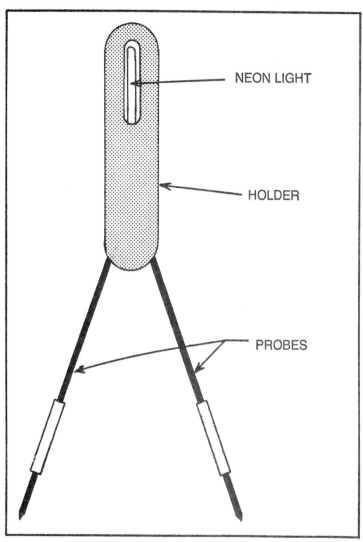

Fig. 1-1. An inexpensive voltage indicator that you can purchase is a neon light attached to two probes. When the probes are placed on live terminals, the neon light will glow. These testers are usually available in two voltage ranges: one tester tests 12 volts and 24 volts, while another tests 120 volts and 240 volts. When a single tester is used to test voltage across a range of voltages in this manner, the bulb will burn brightly when testing the high voltage, and will glow much less when testing the low voltage.

terminals is not known, set the meter to read the highest voltage scale. Then, if the needle moves too little to get an accurate reading on the large scale, turn the dial down to lower scales. When reading lower scales, the needle will swing farther across the scale to make accurate low-voltage readings. But you must always begin on a high scale when voltage is unknown to prevent damaging the meter.

Damage can also occur to a volt-ohm meter if the meter is set on ohms when checking voltage. **Always check your meter dial before placing the probes on terminals.**

When you are checking voltage and no reading appears, first check your meter to be sure it is on the proper setting. Then recheck the power controls to be sure they are set to deliver power to the component. You can't check the voltage reaching your furnace fan motor terminals if the fan switch is open, because the switch is the control that turns on power to the fan.

Making Your Own Voltage Tester

You can make your own voltage tester that will work much like a neon light tester. All you need is a pigtail light socket, available at almost any hardware store or electrical supply shop, and a 25 watt, 120 volt household light bulb. Fig. 1-4. Cut the insulation off about 3/4 inch from the end of each lead, twist the wires tight, and solder them so they'll hold together strongly. With this simple, inexpensive tester you can check the voltage on any 120-volt circuit. Touch the pigtail leads to the terminals, and if the bulb burns brightly, there is voltage across the terminals. Of course, this type of tester has the same disadvantage as a neon light voltage tester. If you need an accurate voltage measurement, you can't get it with this tester. Figure 1-4 also shows how you can make a 240 volt tester by connecting two pigtail sockets together.

Safety Precautions

To make a voltage check, the power must be turned on. Always treat 120 volt circuits and 240 volt circuits with respect. Each is capable of killing you with a shock. When making electrical checks around hot terminals, wear insulating gloves, or attach the voltmeter leads with power off and turn power on only after you have removed your hands from the area of the hot terminals.

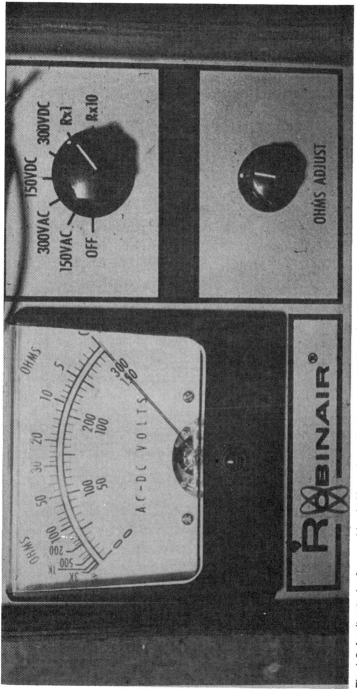

Fig. 1-2. A voltmeter is often combined with an ohmmeter in one meter that can perform double duty. The more expensive meters have a number of scales and will check AC voltage and DC voltage.

OHMMETERS

Ohmmeters are used to determine the amount of electrical resistance between two terminals. In general servicing and troubleshooting, however, precise ohm readings are seldom needed, so the ohmmeter finds its greatest usefulness as a continuity checker. As a continuity tester, the ohmmeter tells you if electricity can flow between two terminals.

Ohmmeters are seldom available as separate meters. Usually they come as part of a volt-ohmmeter, and the user sets the dial to switch the device from volts to ohms (Fig. 1-3).

When checking ohms, the power supply to the component you are checking should be disconnected. The ohmmeter has its own power—a battery or AC circuit that sends electrical current through the circuit being tested. **An ohmmeter will be damaged if it is attached to hot terminals.**

To use the ohmmeter, first set the selector dial for the desired scale. Many ohmmeters have several scales to give more exact readings at various resistance ranges. A typical selector dial might have scales such as R1, R10, and R100 (or X1, X10, and X100). Such settings are easy enough to understand. When the selector dial is set for R10 or a similar setting, multiply the ohm reading on the scale by 10.

With the probes not touching and the meter turned on, the meter will be at the far left of the scale at the *infinity*. Fig. 1-6. Anytime you are using an ohmmeter and get such a

Fig. 1-3. A volt-ohm meter small enough to be hand-held. To use as an ohmmeter, plug in the probes to the body of the meter and, set the dial for ohms.

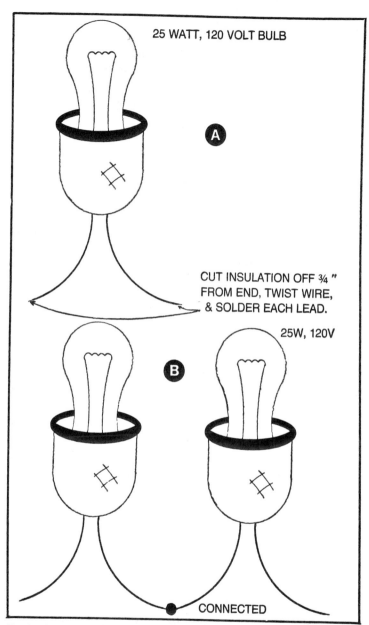

Fig. 1-4. Making your own voltage testers. (A) shows a 120 volt tester made from a pigtail light bulb socket and a 25-watt 120-volt bulb. When the leads are connected to "hot" terminals, the bulb will burn brightly. (B) shows a 240-volt tester made from two pigtail sockets connected together and 2 25-watt bulbs. The 120-volt tester will not work in a 240 volt circuit because the household light bulb will burn out.

reading, it means there is no electrical circuit between the probes. If you were checking an electrical circuit and got such a reading, that would indicate the circuit is *open* somewhere and there is no complete path for electricity to flow.

Now, touch the ohmmeter probes together so there is a complete electrical circuit. The meter will swing across the

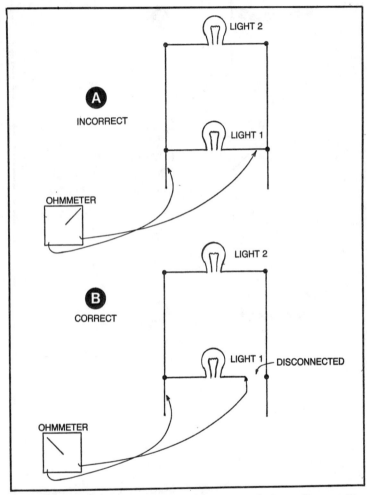

Fig. 1-5. When checking continuity through a component, always disconnect the component from the electrical circuit to prevent "feedback" through other parts of the circuit from giving an incorrect reading. If you are testing light one as shown in (A), even if light one is defective, there is still a complete electrical path through light two, and the ohmmeter indicates continuity. When light one is disconnected and the ohmmeter attached as in (B), only the continuity of light one and its circuit is tested.

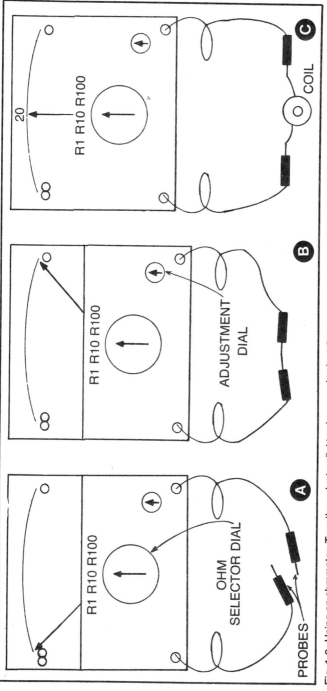

Fig. 1-6. Using an ohmmeter. Turn the selector dial to ohms, and select the range you want to check. Here, R10 has been selected, meaning the correct ohms reading is the figure on the scale multiplied by 10. When the probes are not in contact with a continuous electrical path, the meter reading is infinity, at the far left of the scale, as shown in (A). Touch the probes together, as in (B), to adjust the meter scale to 0. The ohms adjustment dial will set the needle to 0. The ohmmeter is being used to check the resistance through a coil. The meter reading of 20 shows there is continuity through the circuit, but there is also some resistance. The resistance through this coil is 20 × 10 = 200 ohms.

scale to 0 at the far right, indicating no resistance and a complete circuit. The meter has an adjustment dial that should be rotated until the needle points exactly to zero. If you change scales, say from R10 to R100, you will have to readjust this dial.

Checking the Circuit

Now you are ready to check the circuit or component. Place the probes on the terminals to be checked. If you get a zero reading, there is electricity flowing between the terminals. Of course, this may not be the desired result. If the terminals are supposed to be insulated from each other, any reading but infinity indicates an electrical path and a short circuit between the terminals.

Be sure the power is disconnected from the component you are checking and that the component is disconnected from the circuit to prevent "feedback" through other parts of the circuit. See Fig. 1-5.

In most ohmmeter applications, you will find that you do not need an accurate ohm reading. Usually all you need to know is whether there is continuity between two points. An exception to this is when you need to know the resistance in a motor winding or a transformer winding.

Making Your Own Continuity Tester.

A homemade continuity tester will perform the most common function of the ohmmeter, indicating whether electricity is flowing between two points. To make the tester you need a 1½ volt or 6 volt flashlight battery, and a flashlight bulb of the same voltage. Solder electrical wires as shown in Fig. 1-7 to the battery and bulb. Cut off ¾ inch of insulation from the ends of the two wires you will use as probes, twist each wire, and solder to make a strong probe.

When you attach the probes to a closed electrical circuit, the light bulb will glow to indicate continuity. You should remember to check the tester before each test to be sure the battery is still good and the light bulb is good. Touch the probes together. If the bulb glows, the tester is working.

When storing the tester, keep the probes insulated from each other to prevent draining the battery. You can also prevent this by installing a simple low-voltage switch in one of the tester's electrical lines.

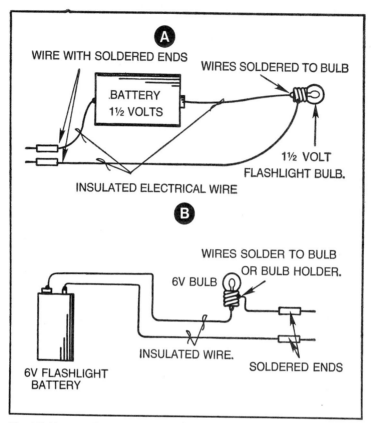

Fig. 1-7. Homemade continuity testers may be made from disgarded flashlight parts. Simply solder an insulated wire to each end of a flashlight battery and solder a flashlight bulb into the circuit as shown. The probes are the loose ends of the electrical wires. Strip the insulation from ¾ inch of the end of each wire, twist each wire and solder it. (A) shows a 1½-volt tester. (B) shows a 6 volt tester.

CLOSED, SHORT AND OPEN CIRCUITS

Every servicing field has its own jargon, and usually it's not necessary to know it to understand how to do servicing. If you understand the parts, and principles, the jargon comes easily once you begin talking with a few "old timers" about the products you are working on. Besides, you often find that different people call the same procedures or parts by different names.

In troubleshooting electrical components, however, it is impossible to avoid discussing closed, open and short circuits. The problem is many people misunderstand what each of these is.

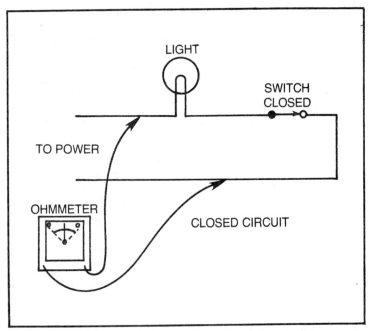

Fig. 1-8. A closed circuit. Electricity can flow. The ohmmeter reading equals the resistance of the light bulb.

Figure 1-8 illustrates a *closed* circuit. This is a continuous connected path through which electricity can flow.

Figure 1-9 illustrates and *open* circuit. When confusion exists, it is most likely to occur in misnaming an open circuit. How many times have you heard people talk of *short* circuits? Probably more than half of them weren't refering to short circuits at all, but to *open* circuits. Open circuits are much more common than short circuits, but many people speak of "having a short", when what they really have is an "open". An open circuit is a path through which electricity normally flows, but which is not continuous because the circuit has been disconnected. Electricity, of course, cannot flow through such a circuit. This break in the circuit can be caused intentionally, such as an open switch or relay, or it may occur through an unintentional break in the electrical wiring.

Figure 1-10 shows a *short* circuit. Short circuits occur when electricity flows through a path other than the one intended. This may happen when a loose or frayed wire touches a metal cabinet, allowing electricity to flow through the cabinet. A common example is a loose wire in an electric motor winding that touches the motor housing. Electricity will flow between the winding terminal and the motor housing.

AMMETERS

It is true that a volt-ohm meter will take care of virtually all your electrical testing needs if you are just beginning fur-

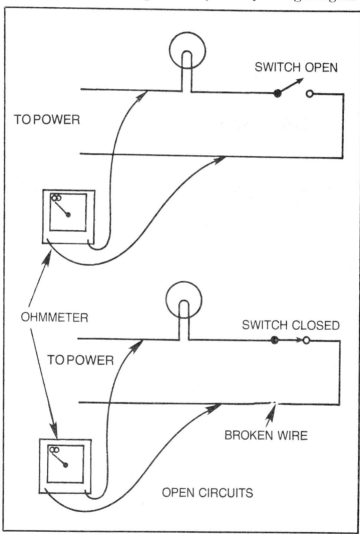

Fig. 1-9. Open circuits occur intentionally, as when a switch is open, and unintentionally, as when a wire is broken. Electricity cannot flow through the intended circuit. Ohmmeter reading: infinity.

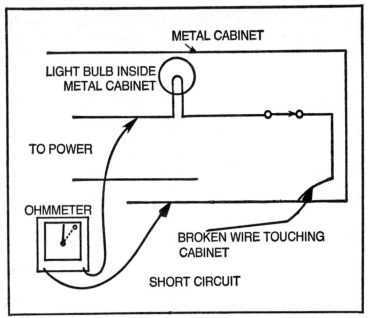

Fig. 1-10. Short circuit is a term often used misused and applied incorrectly in reference to open circuits. In a short circuit, the electricity flows, but through an unintended path. Ohmmeter reading: resistance of path through wire and cabinet.

nace servicing work. If you plan to ever do more extensive work, you'll want to have an ammeter. If you're thinking of doing part time furnace tune-up and servicing either on your own or part of another related servicing business, you might as well consider purchasing an ammeter as one of your first electrical testers. The nice thing about an ammeter is that many of these come in a hand-held unit that is also a voltmeter and an ohmeter. See Fig. 1-12. Thus, for a bit more than you'd pay for a top-quality volt-ohm meter, you will have a volt-ohm-ammeter.

The ammeter measures the current flow through a wire and tells you instantly whether current is reaching the component you're checking. The current is measured in amperes, usually called amps, and is read on the amp scale of the meter. To use the clip-on ammeter—the handiest type—you simply open the upper jaws of the ammeter and clip the jaws so they encircle the wire you are checking. See Fig. 1-13. The component must be turned on and all controls must be set so that electricity will reach the component and flow through the wire you're checking. Notice that you do not have to touch probes to component terminals to make an ammeter check. All you have to do is clip the meter around a wire. If electricity is flowing through the wire, the amperage will register on the meter.

Other types of ammeters besides the clip-on type are available, but they require you to cut a line and attach the meter in series with the circuit. The clip-on, obviously, is considerably handier. **When using a clip-on ammeter, be careful to place only one wire inside the jaws.** If two,

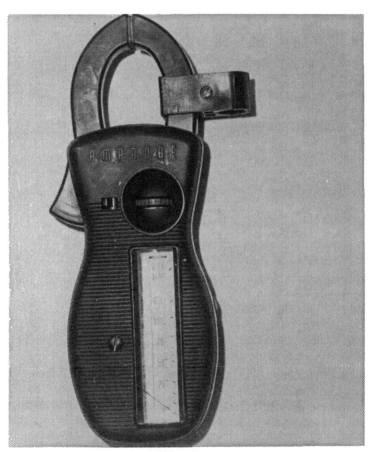

Fig. 1-11. A clip-on ammeter. To check the current passing through a wire, press a lever to open the jaw so that the upper jaws surround the wire. Do not use the jaws to clamp the wire in the jaw tips. A dial rotates the scales to make the scale being used visible in the window. On the side of the meter is an ohms adjustment knob. Be sure to only check one wire at a time.

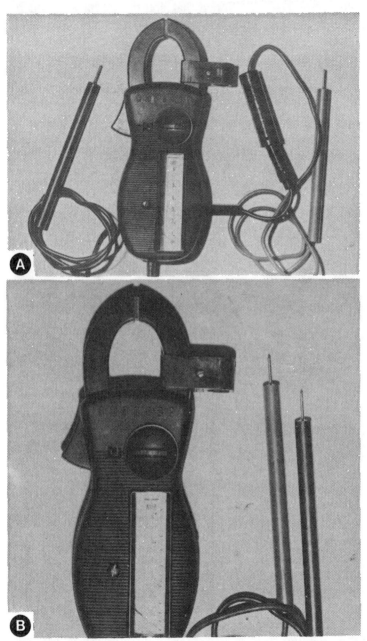

Fig. 1-12A. This clip-on ammeter serves as a voltmeter and an ohmmeter when the probes are connected and the scale is set to read volts or ohms. (A) Probes are connected for the ohmmeter. (B) Probes are connected for the voltmeter.

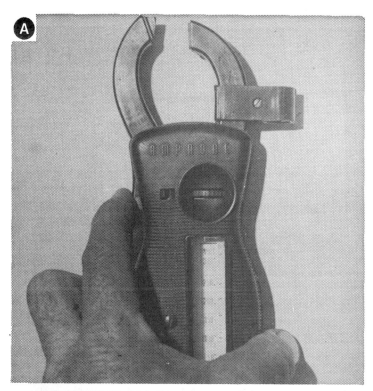

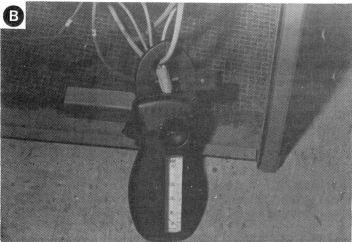

Fig. 1-13. (A) To use the clip-on ammeter, open the ammeter jaws. (B) Use the jaws to encircle the wire being checked. When current is flowing through the wire, the ammeter will show the number of amps flowing. Do not use the ammeter jaws to clamp the wire being tested! Just surround the wire with the ammeter jaws.

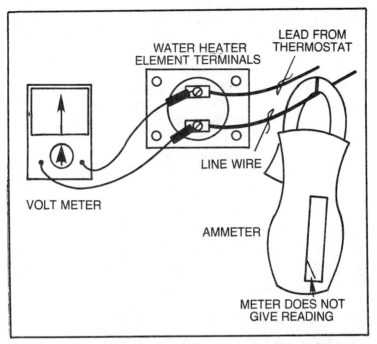

Fig. 1-14. An ammeter is good for checking a defective motor winding, transformer, or heating element that might not show up when you check it with a voltmeter. When you touch the two terminals with the voltmeter probes, the proper voltage reading appears. But the component may not be drawing the correct amount of amperes from the power supply. Here, an ammeter reading taken on an electric water heater element shows no amperage, indicating a defective element. When an electric motor is checked, some amperage will usually be indicated on the meter, but the amperage draw will not be correct for the motor if the motor is defective.

wires are in the jaws, a distorted reading or no reading at all will result.

The ammeter is usually necessary to check a malfunctioning electric motor (such as a fan motor or compressor motor) because a defective motor may appear to be working. In electric motors and some other electrical components, there may be proper voltage to the component, but the component may be defective and drawing the incorrect amount of electrical current. See Fig. 1-14. An ammeter indicates this problem instantly. For instance, if an electric motor is pulling more amps than it is supposed to because of a defective running capacitor, a voltmeter will indicate the proper voltage at the terminals, and the ohmmeter may indicate proper continuity. The ammeter will indicate the excessive amperage draw immediately, and you can turn your attentions to troubleshooting the motor rather than wasting time on an unproductive search elsewhere in the system.

Before making an ammeter check, determine what the amperage through the component is supposed to be. Electric motors should have an amp rating on the nameplate to tell you what this amperage draw should be. Once you know the correct amperage draw, you know which amp scale to set the ammeter for. As with a voltmeter, if you do not know the correct amperage through a wire, begin your check with the ammeter set on the highest scale to avoid damaging the meter. If the amperage turns out to be lower than that range, you can turn the ammeter to a lower scale to get a more exact reading.

When using the clip-on ammeter on a wire with a small amperage draw, you may be able to get a more exact reading on the meter by looping the wire around the ammeter jaws as shown in Fig. 1-15. Divide the reading on the ammeter by the number of loops.

Most ammeters also can be used as voltmeters and ohmmeters when you plug in the leads to the probes and set the meter scale to read volts or ohms. You should follow the manufacturer's directions for your meter to set it up as a voltmeter or an ohmmeter. When using it as an ohmmeter,

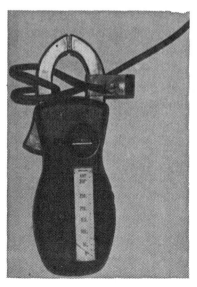

Fig. 1-15. When the amperage draw through a wire is so small that the meter reading is difficult, you can increase the meter reading by looping the wire around the ammeter jaws so there are two or more loops through the jaw enclosure. Divide the reading by the number of loops through the jaws to find the accurate amp flow through the wire.

you should check the batteries periodically, and it's a good idea to carry some spare batteries. It seems your meter's batteries always go dead at the most inconvenient times! The procedures for using the volt-ohm functions are the same for this device as for any other volt-ohm tester. Be sure to start checking on the highest voltage scale when you aren't sure of the voltage to avoid meter damage.

What You Should Know About Your Heating System

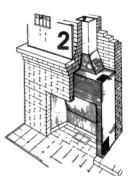

Most heating systems installed today are central heating systems, which means a single furnace supplies heat to all, or most of the rooms in a home, rather than each room having an individual heat source. The heat produced at the furnace is forced to the rooms through a duct system connecting the rooms and the furnace as shown in Fig. 2-1.

Central heating systems generally use one of three fuels: electricity, gas (either liquid petroleum or natural gas) or oil. There are, of course, other fuels and other heating systems. Wood is a popular fuel, especially in fireplaces, as a supplemental source of heat. Many older homes and apartments and some new homes have hot water or steam heating systems. These systems use a fuel-powered boiler to heat water and send the hot water or steam through pipes into the rooms.

Furnaces and other heating systems are normally rated in British Thermal Units (BTU). The BTU rating tells you how much heat your heating system is capable of producing. To get a heating system which is the correct capacity for your house, you match the heating system BTU rating with the *heat load* of the house. By using a set of charts and formulas, a heating technician can tell you the heat load of your house. The heat load actually tells you how many BTU of heat will flow out of the house on a given day. Whatever this heat loss is, the furnace must generate enough heat to replace it and maintain the room temperature inside. Your home's heat load depends on your climate, the size of the house, insulation and the amount of window area.

Electric heating units are sometimes rated in kilowatts instead of BTU. You can convert the kilowatt rating to BTU at the rate of 3.413 BTU=1 watt.

The remainder of the chapter discusses the operation of the various types of heating systems commonly used. Since central heating furnaces are so common, we will first discuss the operation of those systems. Many of the basic principles of the central furnace are also found in other systems.

AIR FLOW THROUGH THE FURNACE

When your furnace heats your home, it draws air from inside the house and brings it to the furnace where the heating units warm the air. Then the air is returned through the duct system to the living area.

A single fan is usually sufficient in a residential heating system to move all the air through the home. The fan pulls the air into the system, pushes it through the heat exchanger and then pushes the warm air through the duct system into the rooms. Obviously, the fan does a lot of work. If it is not the correct size, you will find out soon enough because your home will be cold.

At the heart of the furnace is the heat chamber. In the heat chamber the fuel burns or the electrical elements are heated to warm the air. Actually, the heat chamber is composed of two components: the combustion chamber and the heat exchanger. In a gas or oil furnace the fuel ignites inside the combustion chamber, warming the walls of the heat exchanger above. Air passing through the heat exchanger picks up heat and carries it into the living area. The burning fuel is sealed from the air entering the house. The electric furnace does not need a combustion chamber, so the heat chamber consists of banks of electric resistance elements that warm passing air. On an electric furnace, this is also called the heat exchanger.

Air in a central forced-air furnace moves from the fan past the heat chamber, as shown in Fig. 2-2. In an oil or a gas furnace where a fuel is being burned to produce the heat, the air flowing through the furnace does not come in contact with

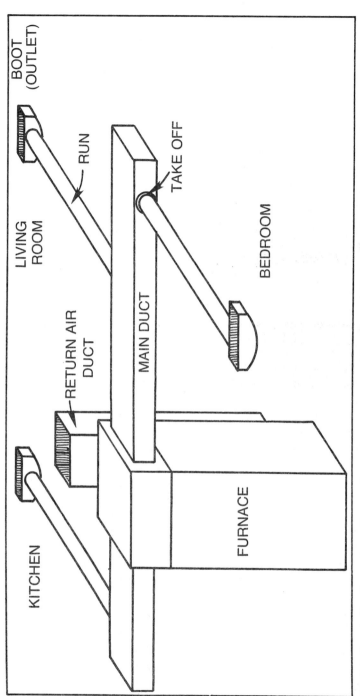

Fig. 2-1. A central heating system relies on a furnace to produce enough heat for an entire home and a duct system to take the warm air to each room. Each room in the home has at least one duct, known as a "run" to deliver air to the room.

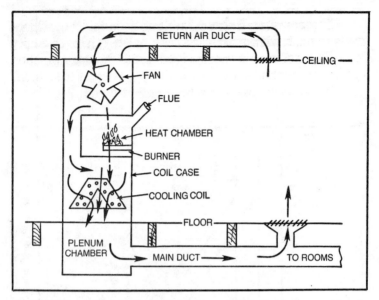

Fig. 2-2. This diagram shows the parts of a central heating furnace and the air flow through the system. This furnace is a downflow furnace, with the air flowing from the top of the furnace to the plenum chamber at the bottom.

the burners. That would put a lot of smoke into the house. Instead, the burners are inside a combustion chamber and are sealed from the air moving through the furnace and into the house. Combustion warms the walls of the heat exchanger, and air moving past these walls picks up heat and takes the heat into the home. To transfer the most heat possible, the heat exchanger is constructed with hollow cores as shown in Fig. 2-3. These hollow cores run the length of the heat chamber and room air passes through them to pick up heat. A flue connected to the top of the heat chamber removes the burned gases and takes them outside the house.

With an electric furnace there is no combustion, so the moving air comes in direct contact with the heating elements. The electric furnace also has no flue. If your home has central air conditioning, the next furnace component the air reaches is the cooling coil.

The cooling coil, also known as the evaporator, is the part of the furnace that provides air conditioning during the summer if you have a combination heating-cooling system. The body of the furnace surrounding the evaporator is the coil case.

If your system does not double as a central air conditioning system, it will not have a coil case, or it may have an empty one. The coil case is not involved in the furnace operation during the heating season since the evaporator is turned off. Figure 2-4 shows the separate furnace components.

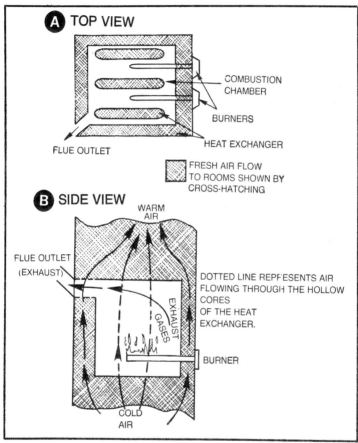

Fig. 2-3. The heat transfer process in a fuel furnace is accomplished through a heat exchanger and a combustion chamber. The top view of the assembly (A) shows the heat exchanger to be constructed with hollow cores passing through the combustion chamber. Air through the furnace flows through the hollow cores and around the sides of the heat exchanger to pick up heat before going to the house. Note that the burners and combustion gases are sealed from the fresh air flow. The burned gases are exhausted through the flue. The heat exchanger seals the room air from the burning flame. (B). The dotted arrows in this side view diagram represent the air flow through the hollow heat exchanger cores that run vertically through the combustion chamber. Burned gases in the combustion chamber are sealed away from the air flow and are expelled through the flue.

The air leaves the furnace from the coil case and enters the first part of the duct system, the plenum chamber. This sheet metal chamber acts as an 18-inch extension on the furnace to connect the furnace to the supply ducts. Depending on whether air flows up or down through the furnace, the plenum will extend through the ceiling or through the floor to make this connection. If the air flow is upward, (upflow furnace) and the furnace is located on the main floor of the house, the plenum extends through the ceiling; if the air flows downward (downflow furnace), it extends through the floor. Air flow direction is discussed later in this chapter.

The main supply duct (also known as the main duct or the main trunk line) carries the air from the plenum chamber to the individual ducts, or runs. Since the main duct supplies all the other ducts, it is the largest duct in the furnace system. Each run is connected to the main duct with a take-off connection. Some furnace duct systems use two main ducts, a frequent installation in long ranch-style homes where the furnace is located near the center of the home. Here, the most even heating is obtained if two main ducts, one in each direction, leave the plenum chamber to supply the runs (Fig. 2-5).

The runs take the air to each room, and the air circulates through the house. To recycle and reheat the air, a return air duct draws air from the living area to return that air to the furnace. The return air duct enters the furnace at the end opposite the plenum chamber. Figure 2-6 shows how the duct system and return air duct are installed in some typical furnace installations. Notice that for the best air circulation, the duct system and return air duct are in opposite horizontal planes (floor or ceiling) from each other.

An important component of air circulation is the furnace's filter. Without a filter to remove dust, dirt and lint from the air passing through the furnace, the furnace would become hopelessly clogged in a matter of weeks. Your furnace will have at least one filter, and it may have more than one. Filters are discussed fully in Chapter 4.

These are the basic components of a central heating system in the order air travels through them: fan, heat chamber, cooling coil, plenum chamber, main supply duct, runs, living area, return air duct and filter. From here the air flow repeats the cycle.

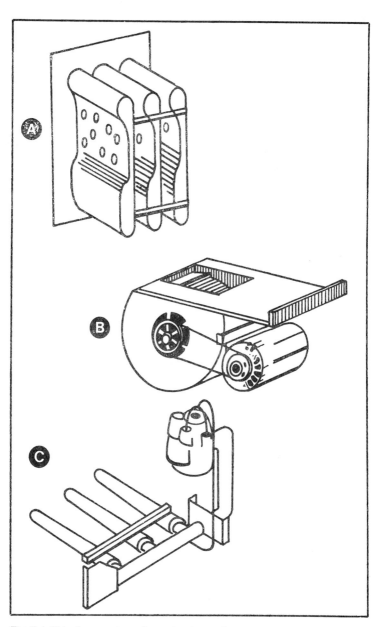

Fig. 2-4. This diagram shows the parts of a gas furnace assembled in an upflow furnace (air flow from bottom to top). (A) The heat exchanger is formed so that burners and combustion gases are sealed off from the air flow. The burners (C) are located at the bottom of the heat exchanger, and heat produced by combustion at the burners warms the heat exchanger walls. Air passing through the heat exchanger picks up this heat and takes it to the rooms. Air is pushed through the furnace and cut system by the squirrel-cage fan (B).

UPFLOW, DOWNFLOW AND HORIZONTAL FURNACES

These terms categorize furnaces according to the direction air flows through them. The upflow furnace is shown in Fig. 2-7. The direction of air flow is from bottom to top, which makes this furnace particularly adaptable for basement furnace installations. The upflow furnace may also be installed on the main level of the home. In this installation the main duct is in the attic above the ceiling, and the return air duct is beneath the floor.

The downflow furnace with the plenum chamber under the floor and the return air duct above the ceiling is a common furnace installation. Air flow is from top to bottom.

The horizontal furnace is shown in Fig. 2-8. This furnace is commonly installed in attics and in crawl spaces in homes where a lack of space in the living area makes the upflow and downflow furnaces impractical. With the horizontal furnace the components are in a horizontal line, and air flows from one end of the furnace to the other.

An important thing to remember is that the direction of the air flow will determine how the furnace parts are stacked on each other. Notice that in all furnaces, upflow, downflow or horizontal, the parts are arranged so air will flow through them in this sequence: return air duct, fan, heat chamber, coil case and plenum chamber.

Except for electric ones, furnaces are purchased to direct the air flow in one particular direction. You cannot switch the arrangement of the parts of an upflow furnace to make it into a downflow furnace, because a specific construction of the furnace parts is necessary for the different air flow directions. A burning fuel has to burn in an upright position.

It is possible, however, as shown in Fig. 2-9 to extend the plenum chamber and the return air duct the height of the furnace to direct the air flow as required for your home. The biggest disadvantage here is that the plenum chamber must be a constant size to get correct air flow through the duct system, and the plenum chamber must run the height of the furnace. That, of course, takes up a good deal of space. But by using such an extended plenum, you can install an upflow furnace of the main floor of your house and still have the duct system under the floor.

Because the electric furnace does not use combustion to produce heat, the parts of an electric furnace may be rear-

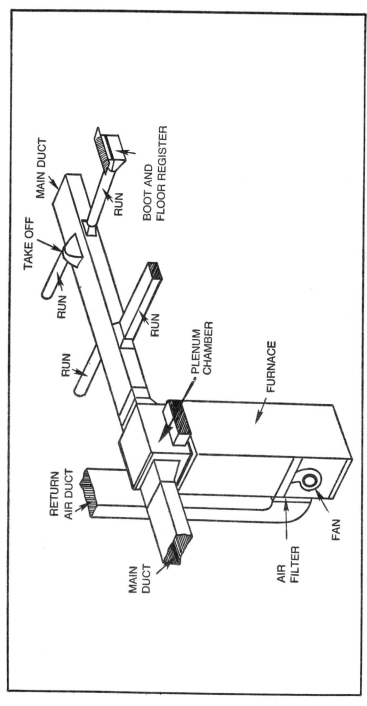

Fig. 2-5. The basic parts of the duct system are shown here. This system has two main ducts, one leaving the plenum chamber in each direction.

ranged to form an upflow, downflow or a horizontal arrangement. The only requirement is they must be connected in the correct air flow sequence. The furnace is installed in the upright position for an upflow or turned upside down for a downflow. It can be laid down for a horizontal furnace.

HEATING-COOLING FURNACES

As we briefly noted earlier in discussing air flow through the furnace, many furnaces are equipped to do double duty. They not only heat during the winter, but they also cool during the summer. This technically makes them "year around temperature control systems", but for brevity we will call them heating-cooling furnaces.

Most modern furnaces today are installed with the capability to serve both functions. The advantages of such an arrangement are apparent. Rather than having a heating-only furnace and several room air conditioners throughout the house, the heating-cooling furnace provides central air conditioning during the summer that is distributed through the home in the same ducts that distribute heat in the winter.

The evaporator coil is the only additional component in the furnace. The evaporator is connected to a condensing unit outside the house that dispenses the heat taken from inside the house. With the evaporator in place and the air conditioning system turned on, the evaporator removes heat from the air as it moves through the furnace.

A central air conditioning system uses the same furnace components in the summer that the heating system uses in the winter. In the summer, however, the combustion chamber is turned off and the fan pushes air past the cooling coil, where the air is cooled, and then into the duct system. The cooling system, of course, is turned off during the heating season.

Even if you do not have central air conditioning, if your furnace was installed within the last few years, it probably has a coil case. The coil case is the portion of the furnace's body that houses an evaporator coil. Most new heating systems are being installed with a coil case so that adding central air conditioning is a simple matter of installing the evaporator inside the coil case and connecting it to a condenser installed outdoors. If your furnace has a coil case but you do not have central air conditioning, there will be an 18-inch empty space

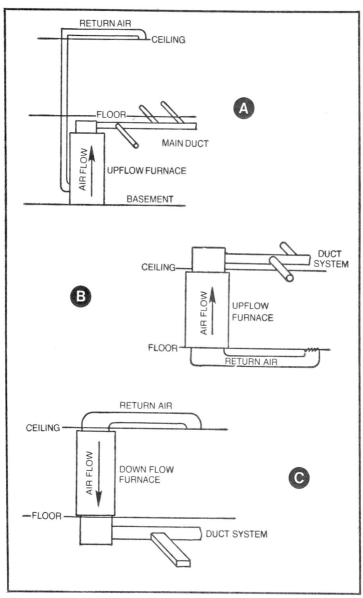

Fig. 2-6. This diagram illustrates three of the most common furnace installations and the positions of the furnaces, duct systems and return air ducts in each. (A) is an upflow furnace located in a basement. The duct system is under the floor and the return air duct runs from the ceiling through a closet or false wall to the bottom of the furnace. (B) is an upflow furnace located on the house's main level. The duct system is above the ceiling and the return air duct is under the floor. (C) is a downflow furnace located on the house's main level. The duct system is below the floor and the return air duct is above the ceiling.

41

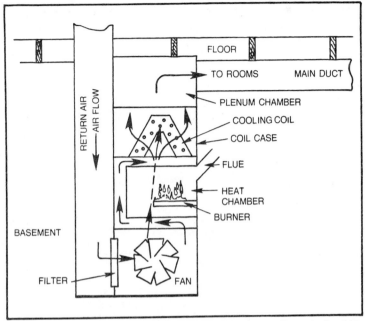

Fig. 2-7. A diagram of the air flow through an upflow furnace. The order in which air flows through an upflow furance's components is the same as for a downflow furnace. The only difference is that in an upflow furnace, air flows from bottom to top.

above or below the heat chamber, depending on the direction air flows through the furnace.

The heating-cooling furnace has a slightly larger fan motor than the heating - only furnace, since cool air is more difficult to circulate than warm air. The heating-cooling furnace uses a different thermostat, and the duct system usually must be insulated to circulate the cool air. Besides these differences, the central heating-cooling system is virtually identical to a heating-only system.

THERMOSTATS

One important part of the modern furnace is the thermostat, although it is not a part of the furnace assembly. The thermostat is the control that turns the furnace on and off, and it is located in the living area of the house where it can be activated by the living area temperature changes.

The role of the thermostat, viewed in its simplest terms, is like that of a light switch: it turns the furnace on, and it turns

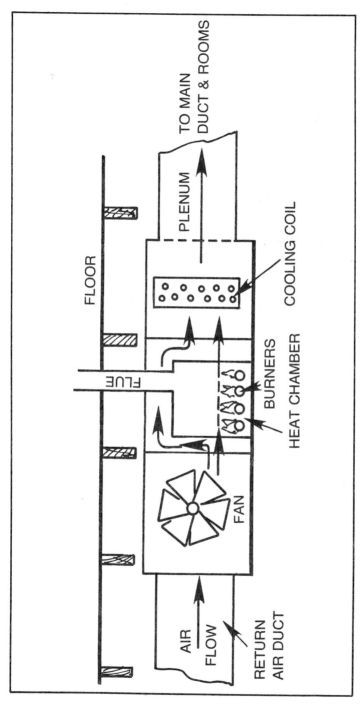

Fig. 2-8. Air flow through a horizontal furnace. Again, air flows through the furnace's components in the same order, but the parts are arranged in a different fashion. The horizontal furnace is a popular furnace for installation in a crawl space or in an attic where height limitations are a factor.

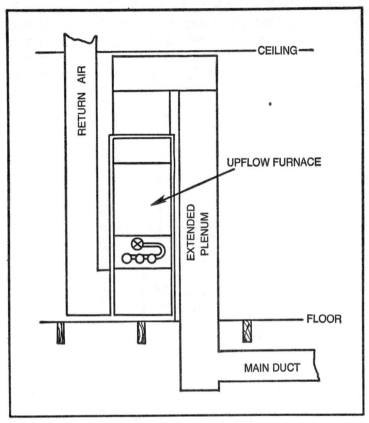

Fig. 2-9. An upflow furnace can be installed on the main floor of a house, even if the duct system is under the floor. Both the plenum chamber and the return air duct must be extended the height of the furnace, which takes up a good deal of space.

it off. Of course, the thermostat is much more complicated than a simple light switch because the thermostat relys on no person to flick a switch to turn the furnace on. The thermostat does that automatically. The homeowner decides at what temperature he wants to keep the house, he sets the thermostat to that temperature, and the thermostat turns the furnace on and off to maintain the temperature at that level.

Thermostats operate with heat sensitive metal contacts that move as the temperature changes. As the temperature of the room drops, the metal will move enough to touch a contact in the thermostat control.

When the contact is made, an electrical connection is completed and low voltage electrical current (usually 24 volts) flows to the furnace, where the electrical current sets in motion the processes that will turn on the heat in the furnace. The heating cycle procedures of different types of furnaces are discussed later in this chapter in the section on furnaces and fuels.

When the furnace has been on and the house has warmed up, the heat-sensitive metal moves again to break the contact and open the electrical circuit. This cuts off the furnace heat. Thermostats are discussed in detail in Chapter 11.

SAFETY CONTROLS

Modern central heating systems have safety controls that shut the furnace off if it begins to overheat. Overheating may be caused by a number of problems, but it commonly occurs because the fan does not turn on to transfer heat from the heat chamber to the rooms. Without a safety control, the temperature in the heat chamber would continue climbing and perhaps cause a fire. The thermostat would not turn off the heating units in this instance because it turns off the heat only after the rooms have been warmed by the circulating air.

Different types of furnaces rely on similar devices, known as limit switches, to protect the furnace and the home from such overheating. When the limit switch located near the heat chamber and flue senses that the temperature has exceeded the preset safety level, it turns off the heating mechanism.

Gas and oil furnaces have limit switches that shut off the flow of fuel to the burners when activated. Sometimes these systems may have two limit controls for safety.

The electric furnace typically has a bimetal switch that turns off the electrical heating elements when the temperature gets too high. A metal disc that melts at high temperatures may also serve as this control.

A wood or coal heating system typically will not have such a safety control. If the stove is automatically controlled from a thermostat mounted on the unit's cabinet, the thermostat will close the damper when the stove reaches a high temperature. Many wood and coal heating systems still used in homes today are small hand-fed systems, and these have no automatic

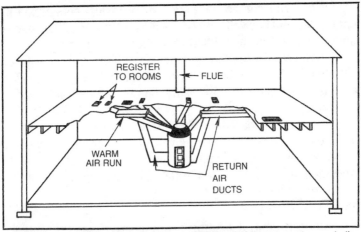

Fig. 2-10. The wood burning furnace is usually a good deal larger than a similar capacity gas or electric furnace. Many of these older style wood furnaces use a "radial" duct design where each individual duct run connects to the plenum chamber at the furnace instead of using a main duct to feed several runs. Often, these older wood furnaces do not have fans to circulate the air, but rely instead on natural convection air currents. If no fan is used to force air through the system, the ducts will be quite large to ease the air flow.

control to stop an excessively hot fire. For this reason, you must be very careful to select a well-constructed stove that will withstand high temperatures. You must also install such a stove so high temperatures pose no danger to the home. These topics are discussed in Chapter 9.

FURNACE OPERATION

Nearly all American homes today are heated with one or more of the following energy sources: wood, coal, natural gas, liquid petroleum (LP) gas, oil and electricity. All these fuels except electricity burn inside the furnace to produce heat.

Wood and Coal

Most residential furnaces installed early in the 20th century burned coal or wood, and many of these furnaces remain in use today. Several companies continue to sell these furnaces, and they are available both in utility and decorator designs.

Wood-burning furnaces are usually among the largest furnaces in size because it takes a large unit to hold enough firewood to burn for eight to twelve hours.

The duct systems for these wood furnaces are often large diameter, short length with few turns and elbows. Figure 2-10 shows a wood-burning furnace design.

The furnace shown in Figs. 2-10 and 2-11, like many older-style wood and coal furnaces, does not have a fan to circulate the air. Thus, the large ducts are used. Instead of a fan, these furnaces rely on the "hot air rises" principle to move the hot air through the duct system.

Without a fan, the heat ducts must come from the house's basement up between the wall panels to a wall register located high on the inside walls. As the air cools it falls to the floor and into the return air duct.

Early coal furnaces had a similar design, but they were somewhat smaller since coal does not require as much storage room as wood. Controls on these early furnaces were entirely manual. Adjusting a dial opened and closed the damper to allow more or less air into the burning fuel.

The addition of a fan to circulate the air through the heating and a thermostatic control system removed the wood or coal furnace's reliance on natural air currents to distribute the heat (Fig. 2-12). When the room temperature dropped below a preset level, the damper would be opened and the heat in the plenum chamber would close the fan control. The fan would run, moving heat into the living area.

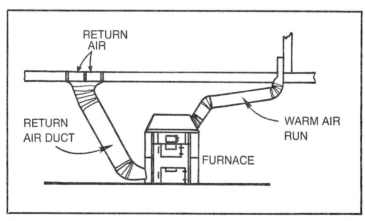

Fig. 2-11. This is a typical design of an older wood or coal furnace. Since the furnace has no fan to force the air through the duct system, it relies on the natural tendency of warm air to rise and cold air to fall to produce the necessary air circulation. The runs and the return air duct are large in diameter and as short as possible. The furnace must be located in the house's basement.

Most modern wood and coal furnaces sold today have fans to circulate the air through the duct system. Even stoves without duct system connections often have auxiliary fans to help circulate the air. Many of these furnaces will burn either wood or coal, provided the correct grate is used.

Natural Gas and LP Gas Furnaces

These furnaces are among the most popular types for residential heating systems. The furnaces are compact and can be installed virtually anywhere in the home. They burn "cleanly", and with proper application there should never be any smoke from the furnace entering the living area.

These furnaces burn two slightly different fuels, natural gas and liquid petroleum (LP) gas. Basically, the furnaces are the same, and you can convert a gas furnace to burn either fuel by changing the gas valves and orifices, the devices that meter the gas into the burners. Other than this difference and the methods of storing the fuels, these furnaces are the same.

Natural gas is a fuel found underground, usually in connection with oil exploration. The gas is placed under pressure and pumped through a series of pipelines crisscrossing the country to take the gas from natural gas fields (usually in the Southwest) to cities where the gas is distributed. Distributors are located in virtually all cities to distribute the gas through underground pipelines to natural gas users.

Liquid petroleum gas, on the other hand, is a fuel that is refined from crude oil and is distributed to users in pressurized transport vehicles. LP gas refers to propane, butane, or a blend of the two. An LP gas customer, unlike a natural gas customer, must have a tank near his home to store the fuel.

Liquid petroleum is stored as a liquid, but the furnace burns a vapor that "boils off" the gas at the boiling temperature of $-50°F$. Under pressure in the storage tank, most of the LP fuel remains a liquid. But some of the space in the tank is occupied by the vapor, which flows to the furnace as the furnace requires fuel. This, in turn, lowers the tank's pressure, and more of the liquid changes to vapor.

Whether you have a natural gas furnace or an LP gas furnace, the gas flows through some sort of tubing to reach the furnace's gas valve. The gas valve opens and closes the flow of

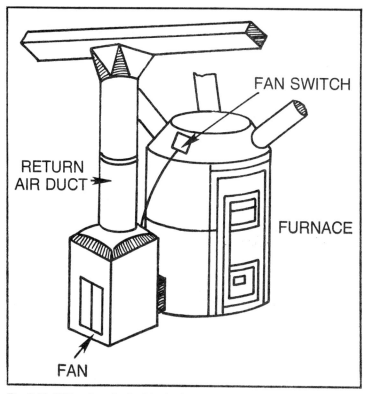

Fig. 2-12. With a fan attached to the furnace, the wood and coal furnace no longer requires natural convection air currents to distribute the heat. A fan control turns the fan on to move air through the return air duct, forcing it through the furnace and into the rooms.

gas to the furnace and, through adjustment, sets the amount of gas that flows to the burners.

As a brief, simple overview, this is how a gas furnace works. The thermostat calls for heat and opens the gas valve. Gas mixes with air as it flows into the burners inside the heat chamber, where the gas is ignited by a pilot light. When a sufficient amount of heat has been generated inside the heat chamber, a switch turns on the fan to begin delivering the heat to the rooms. The burners keep burning and the fan blows until the thermostat senses that the room temperature is high enough. The gas valve is closed, the burners go out and the fan shuts off.

Gas furnaces, their operation, parts and maintenance are discussed at length in Chapter 5.

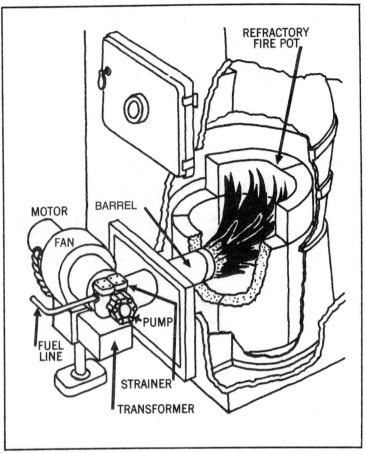

Fig. 2-13. A gun-type oil furnace. A pump brings the oil from the tank and forces it through the air tube (or barrel). At the end of the barrel the nozzles spray the oil and mix it with air. The mixture is ignited by electrodes, and the fire burns inside the brick-lined heat chamber. (courtesy of U.S. Department of Agriculture).

Oil Furnaces

Many oil furnaces burn what is commonly known as diesel fuel, the fuel burned in diesel engines of large transport trucks. Diesel fuel is number two fuel oil, but some oil furnaces also burn other fuel-oil grades or mixtures of similar petroleum distillates.

With oil furnaces, one of the largest problems is handling the fuel. The fuel oil must be stored near the furnace in a tank, but if that tank is outdoors it can cause some problems in extremely cold temperatures. Heating oil, especially the heavier varieties, can turn to a slow-pouring sludge when the temperatures hit bottom.

The most popular oil furnace for homes is the gun type of oil burner (Fig. 2-13). A pump controlled by the thermostat, brings oil from the tank and sends it into the burners within a firebox lined with fire bricks. The nozzles inside the burners spray a fine mist of the oil across a high-voltage electrical arc, which ignites the fuel (Fig. 2-14). The pump, you should notice, must place the oil under a pressure high enough to create the required fine oil mist.

The electrodes serve the same function in an oil furnace that spark plugs do in a car. They must be in adjustment for proper burning. As with a gas furnace, controls turn the fan on when the heat chamber is warm enough, and the thermostat turns the pump off to shut down the furnace.

A second type of oil furnace is the pot type furnace (Fig. 2-15). In this furnace, oil is brought into a pot holding a pool of oil inside the furnace. The heat inside the furnace turns the oil in the pot into an oil vapor that mixes with air to burn. The amount of oil in the pot is controlled by a valve and a float system.

One particular problem of oil furnaces is the soot formed during the burning process. For correct furnace operation, the

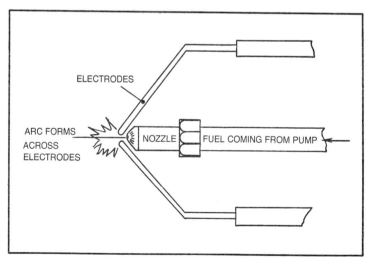

Fig. 2-14. An oil furnace's electrode and nozzle assembly ignites the fuel as it is sprayed through an electrical arc. The nozzle sprays the fuel in a fine mist.

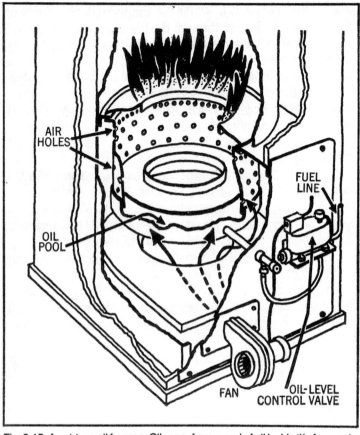

Fig. 2-15. A pot-type oil furnace. Oil vapor from a pool of oil inside the furnace is burned to produce heat. A valve maintains the oil level (courtesy U.S. Department of Agriculture.)

collected soot must be removed regularly from the furnace parts and the flue. Figure 2-16 shows the components of an oil furnace.

Electric Heat

Electric heat does not rely on any fuel combustion to deliver warm air to the home. Because of this, the heat chamber in an electric furnace is nearly maintenance free, and no heat escapes up a flue.

Using the principle that resistance to electric current produces heat, an electric furnace uses banks of electricity resisting metal elements. Thus, electrical heat is called resistance heat (Fig. 2-17). When the thermostat calls for heat, electric current is sent through the heating elements. Heat is produced, which warms the air inside the furnace. The fan carries the heat into the rooms.

Because it has no combustion process and needs only an electrical current to begin operation, electricity is very popular as a fuel for single-room heaters and space heaters. Portable electrical heaters plug into an an ordinary wall outlet and may use a fan to help distribute the heat.

Another popular use for electric heat is in permanent single-room heating installations, such as baseboard heat, ceiling cable and wall heaters. Baseboard heating uses a resistance heating unit installed along a room's baseboard. The heater may be the room's primary or secondary source of heat. Ceiling cable uses resistance heating cord installed in the ceiling to warm a room. Wall heaters use electric resistance elements in a wall-mounted unit.

One advantage to single-room heating units is that each room is controlled with its own thermostat. This allows you to turn down the temperature in rooms when they are not being used. Single-room heaters also make excellent supplementary heating units.

Combination Furnaces

Some furnaces burn a combination of fuels, generally wood and some other combustible fuel. Because wood is one

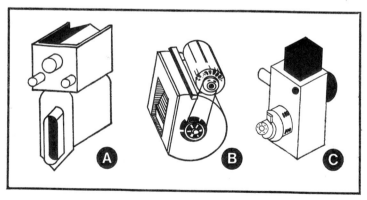

Fig. 2-16. An upflow oil furnace. (A) Heat exchanger. (B) Gun-type burner assembly. Note that the barrel of the burner assembly extends inside the lower part of the heat exchanger when the unit is assembled. (C) Squirrel-cage fan or blower.

of their main fuels, these furnaces are larger than a typical single-fuel furnace. Wood is the primary fuel, and when the wood fire dies, gas or heating oil burners are turned on.

Combination furnaces are discussed further in Chapter 9.

Heat Pumps

Heat pumps are electric heating and cooling systems, but their basic operation does not depend on electric resistance heating principles, as does an electric furnace. The heat pump works instead on refrigeration principles—like an air conditioner—to heat your home. In fact, during the summer cooling season the heat pump *is* an air conditioner, removing heat from inside the home and depositing it outside. The heat pump system reverses during the heating season, removing heat from the air outside and depositing it inside the home. Supplementary electric resistance heating elements are used only when the refrigeration system cannot deliver sufficient heat to warm the home. Heat pumps, their operation, maintenance and service are discussed in Chapter 10 (Heat Pumps).

The main advantage to heat pumps is their efficiency. They generally deliver about twice as much heat per dollar as an electric furnace. Because of this high efficiency and low operating cost (compared to electric resistance heat), heat pumps have received a lot of attention recently.

HYDRONIC HEATING BASICS

The heating systems we have discussed so far can be categorized as warm air heating systems. All depend on the circulation of warm air to heat the home. Most of the systems discussed can also be called forced air heating systems, because they use a fan to blow the warm air through ducts to heat the home. Wood, coal, gas, oil and electric furnaces utilizing a fan would be classified as forced warm air heating systems.

Now we turn to a different method of transferring heat from the heating plant to the room—circulating water. Hot water and steam heating systems use a boiler powered by wood, coal, gas, oil or electricity to heat the water inside the system. This hot water or steam is pumped through pipes to

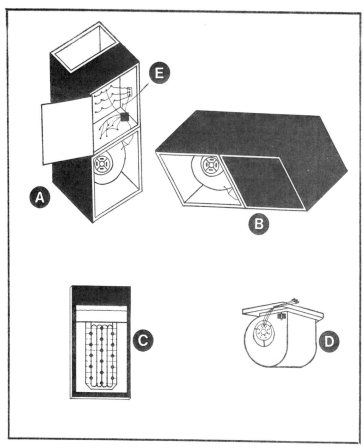

Fig. 2-17. (A) Electric furnace assembled for upflow furnace installation. (B) An electric furnace, unlike gas or oil furnaces, adapts easily to horizontal or downflow installations. (C) Air is warmed by flowing through banks of electric resistance heating elements inside the heat chamber. The heating element terminals are shown at (E). (D) is the fan.

radiators in each room, where the heat is removed from the water and transferred into the living area. The cool water returns to the boiler to be reheated. A diagram of a common forced hot water heating system is shown in Fig. 2-18.

There are several different types of hydronic heating systems available. Hot water heating systems are more popular than steam heating systems because they are generally less expensive and are more responsive to changes in heating demands. Two-pipe hydronic systems are more efficient than single-pipe heating systems, but they are more expensive,

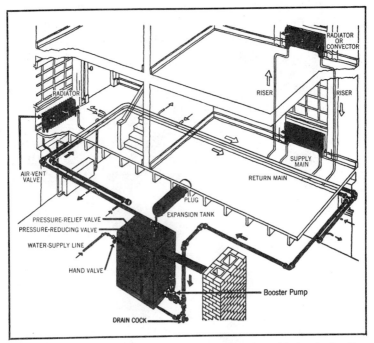

Fig. 2-18. This diagram shows a common two-pipe forced hot water heating system. One water line takes hot water from the boiler to the room heating units (radiators), and the other water line returns cooler water to the boiler. The boiler operates in much the same way a similarly fueled forced-air furnace operates, with the difference that the warm-air furnace transfers combustion heat to air, and the boiler transfers heat to water. (courtesy U.S. Department of Agriculture).

too, because of the extra pipe. In a single-pipe system, the hot water or steam flowing toward the radiators shares the same pipe with the cooler water returning to the boiler.

Forced water systems are more efficient than gravity flow systems. In the forced hot water systems a circulating pump circulates water through the system. The gravity flow system depends on the natural convection currents of warm and cool water to circulate the water through the system.

Boilers

The water flowing through the hydronic heating system is heated in the boiler—the central component of the hydronic heating system. This boiler may be powered by almost any combustible fuel, including all of those mentioned earlier in this chapter as fuels for warm-air furnaces. The three most common fuels for boilers are gas (LP or natural gas), oil, and electricity, but wood and coal-fired boilers are available and are receiving a good deal of attention lately with rising prices of conventional fuels. Figure 2-19 shows an oil-fired boiler and Fig. 2-20 shows a gas-fired boiler.

The combustion process of a boiler is much like the combustion process of a warm-air furnace. Heat is generated inside the combustion chamber, and this heat warms the water inside the boiler. The water, of course, must be sealed out of the combustion chamber. As it heats, the water moves to the top of the boiler, where it is forced through steel or copper pipes that take water to the room registers or radiators.

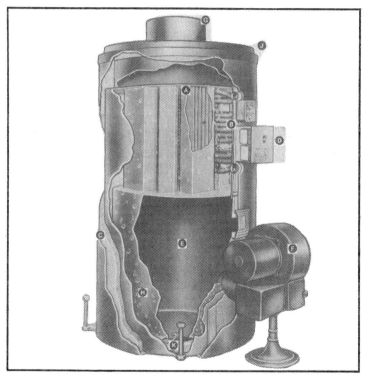

Fig. 2-19. An Oil-fired boiler. (A) Fins line flue surface to remove heat from gases escaping up flue. (B) Copper hot water coil to supply domestic hot water. (C) Jacket Insulation. (D) Hydrostat that controls the boiler operation. (E) Combustion chamber that is sealed from water. (F) Oil burner assembly. (G) Flue outlet. (H) "Wet leg"—Water flows around the combustion chamber to warm the water and maintain a low temperature on the outer surface of the jacket. (I) and (K) Mounting assembly. (J) Steel jacket encloses the unit. (courtesy Edwards Engineering Corporation).

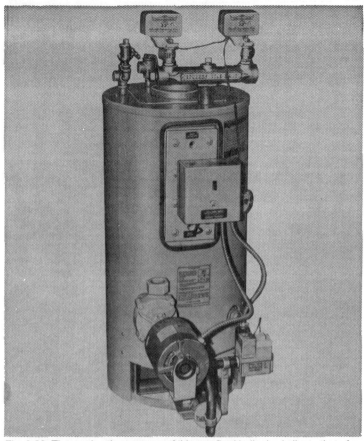

Fig. 2-20. The combustion process of this gas-fired boiler is similar to that of a gas warm-air furnace. (courtesy Edwards Engineering Corporation).

Heat Registers

To transfer the heat from the water into the living area, each room being heated with the hydronic system must have a register. Sometimes these rooms heat registers may be called radiators, but they have come a long way from the unsightly, bulky radiators of years ago. Hydronic heat registers today are compact and unobtrusive. You can purchase baseboard units that fit neatly at the junction of the floor and wall, as shown in Fig. 2-21, or you can purchase wall units that install higher in the wall.

These registers are usually constructed with copper tubing surrounded by fins. The hot water or steam circulates through the tubing, warming the fins. Air circulating through the fins removes this heat flows into the living area. Many room heat registers use natural convection air currents to circulate the air through the register. Since warm air naturally rises, the heated air will rise into the living area, and cool air will replace it in the fins. Some registers, however, use a fan to increase the air flow over the register and transfer more heat into the living area.

When the registers are the forced-air type with fans to move air through the heating unit, summer cooling may be added to the hydronic system. To do this, you must add a water chilling unit to your hydronic system, plus the necessary controls. To be adaptable for summer cooling, the system must have room registers with a drain system to remove water from the registers, and the registers must have fans to circulate the air.

Circulating Pump

Not all hydronic heating systems have circulating pumps to force the water throught the system. In a gravity flow system there is no circulating pump. The hot water rises in the system to room registers, and the cooler water drains back to

Fig. 2-21. This baseboard hydronic heating register transfers heat from the circulating hot water to the living area. Water circulating through the tubing (D) warms the heating fins (E). Air moving between the fins is warmed, and as this warm air rises into the living area it is replaced by cooler air. (A) Cover panel. (B) Damper that adjusts air flow through the register. (C) Support hanger to support the heating element tubing and fins. (F) Plastic strips to prevent metal-to-metal contact when fins expand and contract. (courtesy Edwards Engineering Corporation).

the boiler. Thus, natural convection currents in the water circulate the water.

A circulating pump forces the water through the pipes and registers instead of allowing it to circulate with natural warm water currents. This system is more efficient than the gravity flow system because more water can move faster through the pipes with less heat loss.

The pump is usually located near the boiler. It is controlled by the thermostat, so that when the temperature of the living area drops below a preset level the pump moves the hot water from the boiler into the registers.

Expansion Tank

The expansion tank is attached to the supply main near the boiler, and its purpose is to provide a space half full of air and half full of water to allow room for the water in the system to expand when it is heated.

The expansion tank is sealed so that air cannot escape when the water expands. Expanding water raises the pressure in the system, and as the pressure increases, the boiling point of the water also rises. Through the use of the expansion tank and pressure regulators in hot water systems, hotter water and higher radiator temperatures can be achieved without steam.

Air Bleeder Valve

At some point of relatively high elevation in the hydronic system, there should be an automatic air bleeder valve to remove trapped air from the heating system. Such a valve may be located near the expansion tank or in the return line near the circulating pump. In addition, there may be manual air valves at high points in the system, such as at a radiator.

Air must be removed from the circulation system of a hydronic heating unit. If enough air gets into the system, it can stop the flow of water and cause a good deal of noise in the heating system.

Temperature Controls

Hydronic system temperature controls may be located in a number of different locations—or in a combination of locations on the same system. A central thermostat may control the entire system. Room thermostats may be attached to the room heating registers to control the temperature of each room individually. Another possibility is what has been called *"zone temperature control"*, where one thermostat controls the heating registers in one temperature zone—the rooms on the north face of the home or the bedrooms, for example. Separate thermostats control the different temperature zones in a home. Each zone has its own water loop that attaches to the main supply line and its own circulating pump to pump water through the registers in that zone.

Also, a single circulating pump can be used. Solenoid valves activated by thermostats are installed at the main supply line to open and close the zone loops.

The ability of the hydronic system to heat different zones at varying temperatures with a single heating unit is one of its biggest advantages. With zone temperature controls you can set the thermostat to deliver more heat to cooler north-facing rooms without overheating the rooms with a southern exposure. You can turn down the heat in bedrooms during the day when they are not in use, and turn them up again at night when you get ready for bed.

Other Controls

There are a number of other valves and controls that may appear in a hydronic system. The combustion system controls are similar to those found in a warm-air furnace combustion system, although systems vary on the type of controls that turn the boiler's combustion system on and off. In some systems the burners will continue operating even though no water is being pumped through the pipes to the registers. In others, when the circulating pump stops, the burners also shut down.

If the water inside the boiler gets too hot, a safety limit control will shut off the burners. If pressure inside the system gets too high, the pressure relief valve opens to release excess pressure and prevent damage to the system.

When hot water is sent to two or more room registers, the water flow must be balanced to insure correct water flow and proper heat reaches each room. Gate valves installed in the pipes can be used to maintain the proper water flow balance or the water pipe can be reduced in size to maintain correct water flow to each register.

Selecting The Lowest Cost Fuel

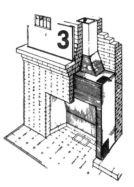

A homeowner purchasing, installing or modifying a heating system today faces a wide assortment of possible heating fuels. Gas, oil and electrically powered furnaces are currently the most popular systems. But coal and wood stoves and furnaces—the most popular heating systems early in this century—are making a strong comeback. In addition, you can choose individual room heaters—either gas or electric—and you can select hydronic heating systems.

It is essential, therefore, to understand a little about each of the possible fuels and heating systems if you are to make a wise selection in modifying an existing furnace to accept a different fuel, adding a second heating source—such as a wood stove—to your home, or in installing a new furnace altogether. Chapter 2 outlined the different types of heating systems available. This chapter discusses how you can decide which fuel is most economical for you—that is, which of the many fuels and heating systems will give you the most heat per dollar.

SQUARE ONE: KNOW YOUR QUESTIONS

Virtually every homeowner wants to heat his house as cheaply as possible. Unfortunately, many don't know how to go about finding out which fuel is most economical for them. Consequently, they go off on a project to change their heating system by adding a fireplace or installing a heat pump, for instance, without any idea just how much they can expect to save in heating costs. The tragedy is that this guessing method usually winds up costing the homeowner money in the long run.

So if you are wondering which fuel is best for you, sit down one evening and figure out just what it is you want from your heating system. And remember that low operating cost is not the only consideration—although it will be the primary focus of this chapter. Availability of supply, ease in handling and expected future costs may lead you to select a more expensive fuel. Some of these considerations are outlined in the previous chapter.

Another thing you must consider in selecting the fuel, whether you are changing a furnace or installing a new one is the initial cost. Some types of heating systems cost more to purchase initially than others, and you must remember that this higher initial cost is an offset against any future fuel savings. Particularly if you are thinking about switching the type of fuel your heating system now uses to another fuel, this is a major consideration.

If you are installing an entirely new furnace system in a new house or to replace an old, worn-out system, your options as to which fuel is most economical are considerably wider than if you already have a working furnace in your home. This is so because if you are installing a heating system from the start, you don't have to worry about what to do with the existing furnace.

If you are considering modifying your heating system, you are probably asking one of the following questions:

- Will I save money by adding a wood stove or fireplace to my home?
- Will I save money by installing a heat pump?
- Since my present fuel is subject to periods of unavailability, how much will it cost to install or add a different-fueled system?
- Will I save operating expenses by installing individual room heaters in some rooms?

FIGURING THE FUEL COSTS

Naturally, the price you pay for a fuel directly affects the overall heating costs of using that fuel. But there are two other factors that must be considered. The first factor is the heat output per unit of fuel. Thus, to compare the heating costs of various fuels, you must know how much heat output each provides. Heat output is measured in BTU (British Thermal Units) for all fuels and heating systems sold in the U.S. Electric units also may be measured in watts at the rate of 1 watt-hour equals 3.413 BTU. Thus, by determining how many BTU a given fuel will produce per fuel unit, you can easily compare the cost per BTU of the fuel and compare that cost to the cost per BTU for all other fuels you are considering. Figure 3-1 shows the BTU output of the common fuels.

The second factor in determining the heating costs of fuels is the heating system efficiency. You must have some

FUEL	UNITS	BTU PER UNIT
HARDWOOD	CORD	20 TO 22 MILLION BTU
SOFTWOOD	CORD	12 TO 14 MILLION BTU
HARD WOOD	1 LB.	8,500 TO 8,600 BTU
COAL	1 TON	20 TO 22 MILLION BTU
	1 LB.	10,000 TO 11,000 BTU
NATURAL GAS	CU. FT	1,000 BTU
	THERM (100 CU FT)	100,000 BTU
PROPANE	GALLON	90,000 BTU
BUTANE	GALLON	130,000 BTU
NO.1 FUEL OIL	GALLON	136,000 BTU
NO. 2 FUEL OIL	GALLON	140,000 BTU
ELECTRICITY	KILOWATT-HOUR	3.413 BTU

Fig. 3-1. This chart shows how many BTU each type of common heating fuel produces per purchasing unit. You should note that this tells you the potential heat available in each unit and includes heat lost in the heating system during combustion that will not actually reach the living area. Also note that propane and butane are gases that are sold and contained in a pressurized form, and they are sold by the gallon. Liquid Petroleum Gas (LPG) is a name commonly given to both, or to a blend of the two. Your LPG fuel supplier should be able to tell you which type of gas his LP gas is and the BTU value per gallon.

idea how much of the heat produced by the fuel goes into the house and how much is wasted in the heating systems you are considering. Only then will you know whether a given fuel-fired heating system will be economical for you.

In comparing the amount of heat per dollar for fuels, what you need to know is how much of the heat produced by the fuels enter the living area. If, for example, you were considering a hand-fed wood stove, it would be important to you that roughly only 25 percent of the heat produced actually enters the living area.

There will be some heat lost in virtually all heating systems. Any fuel-fired heating system—coal, gas, wood, or oil—will have some heat lost up the flue. An electric furnace does not have a combustion process and has no flue, so this heat loss is not present. Still, all furnace and hydronic systems will have some heat loss in the heat distribution system. Such losses can be cut considerably by insulating the furnace ducts or the hot-water pipes to prevent heat loss. The one heating system that will have about 100 percent efficiency is the electric baseboard heater. Virtually all the heat produced by the heating elements ends up in the room because there is no fuel-burning process and no duct system to lose heat. Figure 3-2 shows the efficiencies of most common heating systems.

None of the figures in this chart should be taken as the final word. The efficiency of a heating system depends on numerous factors, and about all we can give you are some rough estimates of what you can reasonably expect the efficiency of a typical heating unit in these categories. You also should be aware that there are conflicting views as to just how the efficiency of a heating system should be measured, and this leads to confusing claims.

Also, you should note that the efficiency figures for heat pump installations will vary widely according to the climate and installation. Chapter 10 presents more exact figures on figuring the efficiency of a heat pump. Likewise, wood-burning units vary widely in their efficiency and are treated later in this chapter.

The Formula

The following formula will tell you how much heat per dollar you will gain from any of the fuels:

$$C = \frac{F \times E}{P}$$

where C = cost per BTU of heat energy, in BTU per dollar,
F = BTU value of fuel per purchasing unit,
E = efficiency of your heating system, in percent,
P = price of fuel in cents per purchasing unit.

Your local fuel suppliers can tell you the cost for one unit of each type of fuel. Remember that some fuel prices are based on *graduated* rates—that is, the cost per unit varies with the amount you purchase. This is a practice of electricity

FUEL & TYPE OF HEATING SYSTEM	EFFICIENCY
GAS FURNACE OR BOILER	60-70%
OIL FURNACE OR BOILER	60-70%
ELECTRIC FURNACE OR BOILER	85-90%
ELECTRIC ROOM HEATING UNITS	95-100%
ELECTRIC HEAT PUMP	120-200%
HAND-FED WOOD STOVE	ABOUT 25%

Fig. 3-2. Different heating systems have different efficiencies, as shown in the chart. In any given installation, the actual efficiency will depend on a variety of factors, but this chart can give you some idea of efficiencies for purposes of comparing fuel costs. Hot water systems and warm air systems have about the same efficiencies for any given fuel. The actual efficiency of any heat pump depends on a wide variety of factors, which are treated later in this chapter and in Chapter 10 (Heat Pumps). Likewise, the actual efficiency of a wood stove will vary greatly according to the way the fire is tended and the wood burned in the fire.

suppliers, in particular, and many other fuel distributors may also give you discounts for the purchase of large enough quantities. For purposes of the formula, the figure you are interested in is average cost per unit over the entire heating season for each fuel. This figure can be obtained by getting an estimate of your heating consumption over a season and the estimated total heating cost; then find the average heating fuel costs. Most fuel suppliers will be able to estimate your fuel requirements and your average unit fuel cost. This cost should include all fuel adjustment factors and taxes.

Some Examples

Suppose you are considering installing a natural gas furnace in your home and that the furnace is 65 percent efficient—a reasonable estimate for most gas furnaces. Assume gas in your area is available for an average cost of 20 cents per *therm* (100 cubic feet). Plugging these figures into the fuel cost formula, you get:

$$C = \frac{100,000 \times 65}{20}$$

$$C = 325,000 \text{ BTU per dollar.}$$

Therefore, for each dollar spent, you would receive 325,000 BTU of heat.

Suppose you are also considering installing a heat pump, with an average seasonal efficiency of 150 percent. If electricity costs 2 cents per kilowatt hour, the result is:

$$C = \frac{3413 \times 150}{2}$$

$$C = 255,975 \text{ BTU per dollar.}$$

Clearly at these prices, a heat pump is not less expensive to operate than a gas furnace. Also, if that gas furnace were tuned up to reach the 75-80 percent level of efficiency, gas heat would cost even less.

If you were considering placing a wood stove in the living room, and wood were available at $50 per cord, you would compare the heat per dollar of wood with the heat per dollar of gas or an electric heat pump to determine whether you would save money. Figuring the heat output of wood adds an additional factor to the calculation, a discount in heat value accounts for the amount of heat absorbed in evaporating water from the wood. This is discussed in the next section, but for now assume 85 percent of the heat potential of the wood under consideration will actually be available for heating. This is the result:

$$C = \frac{22{,}000{,}000 \times 25}{5000} \times .85$$

$$C = 110{,}000 \text{ BTU per dollar} \times .85$$

$$C = 93{,}500 \text{ BTU per dollar}$$

Notice in the formula that the denominator (the price) is always expressed in *cents*, not dollars, so the decimal point can be dropped from the furnace efficiency percentage figure. This saves working with the decimals in the formula. Again, it is clear that gas is the better fuel buy, at these prices.

Wood Heat and Efficency

As mentioned earlier, there are a large number of variables that come into play when you begin considering the heating efficiency and the BTUs per dollar of wood heat. For one thing, it is not nearly so easy to pinpoint the actual efficiency of most wood-burning equipment, and manufacturer's claims are difficult to prove. So much of the efficiency of a wood-burning device depends on how well the fire is tended, what type of wood is burned, etc., that it is much more difficult to say with precision just what the efficiency of a certain type of wood-burning system would be. A few generalizations are possible, however. You can be reasonably certain that the most efficient wood-burning system would be the larger wood-burning furnaces on which air supply is carefully controlled and little fire tending is required. On most of these furnaces, efficiency would approach that of a fuel-fired furnace. That is, they would probably be about 50 to 70 percent efficient.

The second difficulty in determining the heat per dollar for wood-burning installations is that the amount of heat produced varies greatly with the wood you burn. Unlike other fuels, in which the BTU per unit can be pinpointed almost exactly, the BTU of every cord of wood is different.

A cord, of course, is a volume measurement, and it is the standard measurement for most wood sales. The fact is that all wood has about the same heat output per pound—8500 to 8600 BTU. But the woods that we know as softwoods—pine, fir, aspen, etc.,—are less dense and weigh less per cord than the woods we know as hardwoods—oak, hickory, and maple. Therefore, softwoods produce only about 60 percent as much heat as do hardwoods, because they weigh less per cord.

This should make the next point self-evident. When purchasing wood, be sure to discount the value of any softwood you might purchase for the decrease in heat value per cord. Pine wood at $40 per cord is no bargain if oak is $50 per cord. Although the pine is cheaper per cord, it is actually more expensive per BTU, and it is not nearly as desirable a wood to burn, either. Oak scraps from sawmills are not any bargain, either, at $40 per cord if regular oak is $50. Those sawmill scraps usually contain a lot of bark, and bark is lightweight stuff. That means less heat per cord.

If you are buying wood, therefore, know your woods and know the heat you are getting from them. Otherwise, you will be paying more money per BTU in your attempts to save money. Figure 3-3 shows some BTU heat values per cord for some common wood types.

Of course, another factor involved here is how tightly the cord of wood is stacked. Since a cord is simply a volume measurement ($4' \times 4' \times 8'$), there are tightly packed cords and loosely packed ones. The heat values given assume a tightly packed average cord of wood.

Yet another factor involved in determining the amount of heat available in wood is the wood's moisture content. Seasoned wood is generally said to be wood containing about 20 percent moisture. Even this amount of moisture is something that must be evaporated during the burning process, and the heat supplied by the wood must perform this function. The evaporation process absorbs heat—about 15 percent of the heat produced by a seasoned log—so an adjustment factor must be included to account for this heat loss. For seasoned wood, the adjustment factor should be .85 as used in the previous example. Putting this figure in the fuel cost formula

Fig. 3-3. Although all wood has roughly the same amount of heat value per pound, about 8600 BTU, hardwood varieties weigh more and contain more BTU per cord than do softwood varieties. This chart indicates the potential BTU contained in each cord of seasoned wood and does not account for heat that will be lost when moisture is evaporated from the wood. About 15 percent of the amount of heat shown will be lost in this evaporation.

WOOD	BTU PER CORD (IN MILLIONS)
ASH	20
BIRCH	22
ELM	17
HICKORY	25
MAPLE	20
OAK	22
PINE	13

accounts for the heat expended evaporating the wood's moisture in seasoned wood. Of course, if primarily "green" or unseasoned wood were to be burned, a lower factor than .85 would be appropriate.

Wood-Burning Systems

Determining what efficiency figure to plug into the fuel cost formula when you are working with wood-burning systems can be difficult. At best, we can present a few generalizations, and the homeowner must realize that any actual efficiency figures depend greatly on the system design and the way the fire is tended.

A masonry fireplace with a damper but without a heat-circulating fireplace liner or any of the heat-increasing modifications discussed in Chapter 9 will have an efficiency of 0 to 10 percent; a hand-fed wood stove, 20 to 30 percent; a thermostatically controlled stove, 30 to 40 percent; a fireplace with efficiency-increasing modifications, 20 to 30 percent; a wood furnace, 50 to 70 percent.

For most persons planning to use a significant amount of wood to save money on their heating bills, the actual percentages are not really that important. Most of these people have a free source of wood and do not even have to use a fuel cost formula to know that they will save money by burning wood. But if you must purchase your wood supply, these figures can give you a rough idea as to whether wood heat will save you money. These efficiency figures are general only.

FUEL EFFICIENCY AND HEAT PUMPS

The heating efficiency of a heat pump over an entire heating season is measured in terms of a Seasonal Performance Factor (SPF), which is nothing more than a number that tells you, on the average, how much heat a heat pump produces for the electricity input over an entire heating season. A fully efficiency electric room heater is said to have a SPF of 1, meaning it is 100 percent efficient. The SPF of a heat pump, then, tells you how much heat you are obtaining from a heat pump compared to the heat you would get if you put the same amount of electricity into an electric resistance heater. Typical SPFs for heat pumps installed in the U.S. are 1.25 to 2.0, meaning those heat pumps produce 1.25 to 2.0 times the amount of heat for a given amount of electricity as an electric room heater would produce with the same amount. Such heat pumps, as you can see, would be said to have efficiencies of 125 to 200 percent, and these are the percentage numbers you would plug into the fuel cost formula shown earlier to find the BTU per dollar figure for comparison.

Chapter 10 (Heat pumps) contains a section that treats heat pump efficiency and SPFs in detail. Using that section and by talking with a competent heat pump dealer, you can get a rough idea of what the *average* seasonal efficiency of a heat pump would be for your installation. For purposes of working through the fuel cost formula, the SPF figures given in Chapter 10 can easily be translated into efficiency figures—SPF of 1.75 equals 175 percent efficiency.

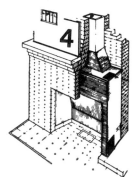

General Furnace Tuneup Tips

Most modern heating systems are designed and built so they will operate for a few years without a bit of servicing or maintenance, but you would be foolish to operate them that way. When a furnace is "out of tune," so to speak, it will operate, but it will not operate properly. It uses more energy, runs more often, works harder, and does not heat as well. If you run it for a long enough period of time without maintenance, you will shorten the furnace's life by several years.

Furnaces, like most mechanical devices, will operate much better when they are in tune. It's really silly not to keep a furnace in tune, because the procedure takes very little of your time. Most routine maintenance is extremely simple. By going over your furnace once a year right before the heating season, you will save energy costs and wear and tear on your furnace.

Unfortunately, many furnaces have been neglected by their homeowners for quite some time. It has only been in recent years that increased concern about rising energy costs has made some homeowners aware of the importance of routine heating system maintenance. If your furnace has been one of the neglected ones, don't worry. Just roll up your sleeves and do a thorough job of cleaning and servicing your furnace as soon as possible. The first time will probably take some time and effort, but when you have to do the job the second time, it will be a snap.

In this chapter we present some of the tune up chores and procedures that need to be done routinely for almost any central furnace heating system. Subsequent chapters describe the tune up and adjustment procedures for specific types of heating systems.

FILTERS

Every furnace has one or more filters to remove dirt, dust and lint from the circulated furnace air. This keeps dirt from entering the furnace and clogging up coils and heat exchangers. Naturally, in a few months these filters become clogged. When this happens, air flow through the furnace and duct system is restricted, greatly reducing the furnace's efficiency. It is very important for furnace efficiency to have free air flow through all parts of the heating system. If a dirty filter continues in use long enough, it will choke off virtually all of the air flow through the furnace. When this happens, it means a great loss in fuel dollars because the heat generated in the furnace stays there instead of being transferred to the living area.

You should clean or replace your furnace filter as part of the annual tune up each year before the start of the heating season. Then, once in about the middle of the heating season, clean the filter again. If you have central air conditioning as well as central heat, you should clean the filter at least once during the cooling season for a total of at least three to four times a year. Until you are familiar with your heating system, you should inspect the furnace filter about once every two months to see if it needs cleaning. Don't hesitate to clean the filter five or six times a year if inspections show large amounts of lint and dirt warrant cleaning it that often. If you have a disposable filter, however, you probably won't want to replace it that often because of the cost of new filters. Replacing a disposable filter once before the heating season and once before the cooling season should be sufficient in most installations. If interim inspections reveal lint or dirt deposits on the filter, you can save money by cleaning the filter as described later and brushing away lint or dirt from the surface of the filter.

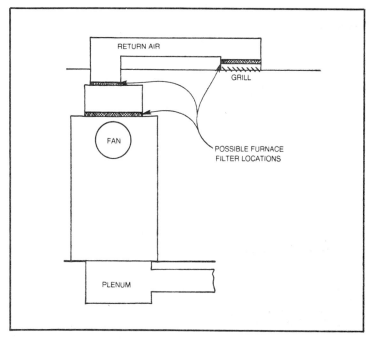

Fig. 4-1. The furnace air filter may be located in any of several locations as indicated, and some furnaces may have more than one filter. Normally these filters are located in the cold air (return air) intake duct before the return air enters the furnace. This filters the air before it goes through the furnace.

Note that the furnace in the diagram is a downflow furnace. If you have an upflow furnace, the return air intake will be at the bottom of the furnace.

Horizontal furnaces are normally installed under the floor of the house or in the attic. Naturally, such an installation makes the filter difficult to reach if it is installed at the furnace. In these furnaces, the filters are often installed at the room return air opening where the filter is much more easily accessible.

Air filters on most furnace installations are accessible and easy to remove. But all furnaces are different, so it's impossible to say with certainty where you will find the filter on your furnace. Figure 4-1 shows some popular locations for furnace filters. Some furnace installations have more than one filter. Installing the air filter behind the return air grill makes the filter accessible for easy cleaning with horizontal furnace installations.

The most common places to find the filter are in the return air duct just before the air enters the furnace.

There are three types of furnace filters: disposable, washable and electronic. Each has different characteristics and maintenance procedures.

Disposable Filters

As their name implies, disposable furnace filters are made to be used once and thrown away when dirty. These filters are normally made of spun glass sandwiched between two thin metal panels and fiber enclosed in a light cardboard frame, as shown in Fig. 4-2. The fiberglass strands are coated with a sticky substance to trap dust in the air.

When you examine a disposable filter, take a look at the sides of the cardboard frame. There will probably be an arrow indicating the direction of air flow through the filter. Make a note of this when you install a disposable filter, and be sure the air will flow in the proper direction through the filter. These air filters are designed so proper air flow and filtering action occur only if the furnace air flows through them in the direction indicated.

You can normally expect to change a disposable filter at the beginning of the heating season and once more during the season. Although disposable filters are made to be thrown away when dirty, you can prolong their life to perhaps as much as an entire heating season by lightly cleaning and dusting them off every month or two. Keeping the filters cleaned between changes means greater air flow through the furnace and better heating efficiency, too.

A disposable filter should never be washed with water or blown out with compressed air. Each will damage the filter beyond repair. To clean a disposable air filter, carefully remove it from the furnace and tap it upside down on a hard surface. This loosens some of the surface dirt and dust particles and knocks them out of the filter. Then, gently vacuum the intake side of the air filter so that the air pulled by the vacuum cleaner comes through the filter in the direction opposite the normal air flow. This procedure should remove more of the loose dirt and dust that the filter has collected.

Of course, nothing short of replacement will remove all the imbedded dirt and lint from the filter, but at least this procedure removes some of the restricting dirt between filter changes. This also keeps your filter cleaner between changes, which increases furnace efficiency and filter life.

Be careful not to bend the cardboard frame or tear the filter element. If this happens, you should replace the filter.

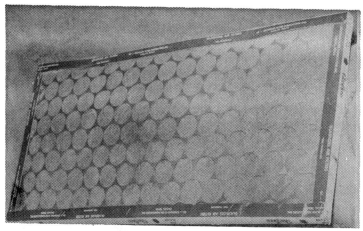

Fig. 4-2. Disposable air filters are constructed of fiberglass strands sandwiched between light metal grills and surrounded by a cardboard frame. The resulting filter is an extremely light unit, but it also is easily damaged from mishandling.

When you replace a disposable filter, be sure to replace it with the *correct size* filter. It is especially important that the replacement filter be the same *thickness* as the old filter. Even though a thicker filter will fit, don't use it! Doing so just causes the very problem you're trying to prevent—restricted air flow through the furnace and reduced furnace efficiency. You can purchase a replacement filter at a heating supply store and in some hardware stores.

Washable Filters

Washable air filters present a couple of advantages over disposable filters. Although their initial cost is higher, they save the homeowner the expense of purchasing replacement filters twice a year or more. They also save him the trouble of locating a replacement filter of the same size.

Washable filters are usually made of a foam material covered with a metal screen and enclosed in the metal frame. The entire frame and filter assembly is easy to locate and remove in most furnaces.

At least once before the heating season begins and once more during the heating season, you should remove the washable filter and clean it thoroughly with soap and warm water. If your inspection reveals that more frequent washings are desirable to keep the filter clean, by all means, wash the filter

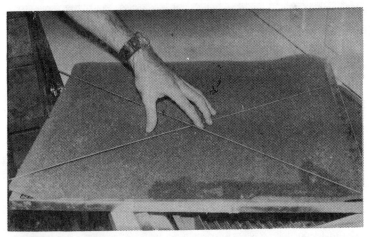

Fig. 4-3. To clean a washable filter, remove the filter element and metal assembly from the furnace. This filter element is held in place by two crossing metal rods that are easy to remove.

more often. Removing the dirt and lint from the filter increases air flow and furnace efficiency.

To clean the washable air filter, you should remove the foam element from its frame as shown in Fig. 4-3 and wash the element gently with soap and water. Run the water through the element in the direction opposite the air flow to remove trapped dirt and lint. See Fig. 4-4.

You shouldn't have to replace a washable filter element unless the element becomes torn or so dirty that you cannot clean it properly. Replacement elements may be available in precut sizes, or you may have to cut the proper size from a large sheet of element material available at a furnace supply shop. Try to get a replacement element that matches the old element exactly, if possible. If the replacement element material is too dense, it will unduly restrict the air flow through the unit. If it is too light, it will allow a large number of dust and dirt particles through. Be sure the replacement element fully fills the metal frame. If it doesn't, air leaks around the edges will allow unfiltered air to pass through.

Electronic Air Filters

Electronic air filters are the most sophisticated of the air filter types currently available for home heating installations. Unlike disposable filters and washable filters, which trap dust and dirt particles only if they are large enough to get caught in the filter webbing, electronic air filters use electronic charges to trap even microscopic particles present in the air. They will trap almost all dust and pollen particles, and for this reason, electronic air filters are favorites of persons who suffer from allergies and dust irritation (Fig. 4-5).

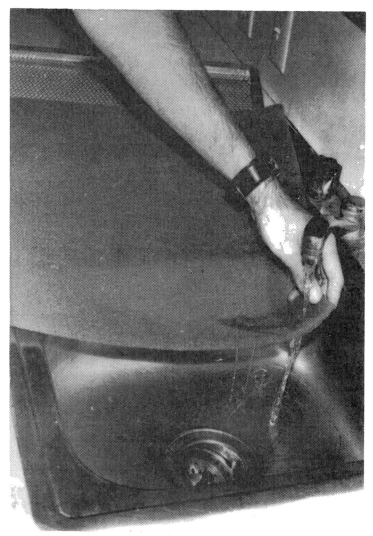

Fig. 4-4. To clean a washable filter, simply run water through the foam filter element in the direction opposite air flow. If needed, soapy water may be used for cleaning, but be sure all the soap gets rinsed out.

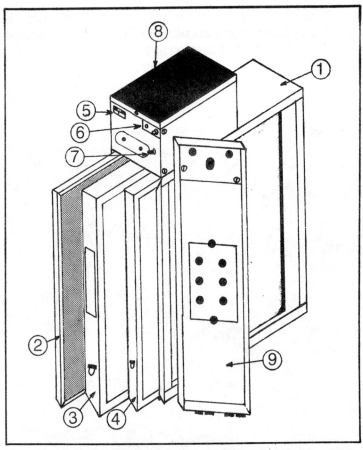

Fig. 4-5. An electronic air filter with cover removed to reveal parts removable for cleaning. (1) Filter cabinet. (2) First stage filter that removes largest dirt and dust particles. (3) Electronic air cleaning cell (4) After filter that collects any particles that may flake off from the electronic plates. (5) Indicator that tells you when the plates need cleaning. (6) Power switch. (7) Door switch that turns power off when door is opened. (8) Power box containing transformers and electrical connections. (9) Removable door.

Electronic air filters are available as options from most furnace and heat pump manufacturers on many models. Most electronic air filters can be found inside the furnace cabinet. The filter elements are removable for cleaning.

The electronic air filter actually cleans the air in a two-step process illustrated in Fig. 4-6. First the air passes through a conventional filter element that traps the larger dirt and lint particles. The remaining particles in the air are passed through charged plates that attract the particles.

The electronic air filter should be cleaned according to the manufacturer's recommendations. Normally, this involves cleaning the electrically charged plates with a special cleaning fluid that removes the collected dust and dirt. The lint screen air filter usually is cleaned in the same manner you would clean a disposable or a washable air filter. Electronic filters normally have an indicator on the outside of the cabinet to tell you when the elements should be cleaned.

Electronic air filters operate using as much as 12,000 volts of power to charge the plates. This much voltage can give you quite a shock if you should happen to touch the plates when the power is turned on. Be certain the power is disconnected. Most units have a switch that automatically disconnects the power supply when the servicing door is opened, but double-check and be sure the power is not on before you begin servicing.

CLEANING THE FURNACE FAN

As part of your regular furnace maintenance program, you should clean, oil and inspect your fan to insure its continued operation. A once-a-year inspection and maintenance can go a long way toward lengthening the life of your fan and

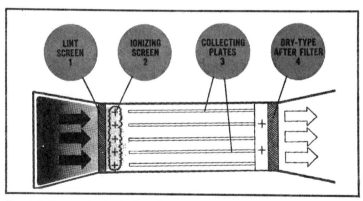

Fig. 4-6. The electronic air filter cleans the air in a two-step process. The air entering the filter unit first passes through a washable lint screen that removes the largest particles of dust from the air. Then, the air passes through an ionizing zone, where the microscopic dust and pollen particles are given an electronic charge. The air goes through the electronic collecting plates next, and the high-voltage charge on these plates attracts the ionized dust particles out of the air, much like static electricity charges will attract certain particles. The voltage on these collecting plates may be 10,000 volts or more. The "after-filter" is the last filtering element. It collects flakes of dirt and dust that eventually will break away from the electrical plate surfaces (courtesy Williamson Company).

Fig. 4-7. The belt-driven squirrel cage fan has an electric motor mounted outside the fan cage, and the motor is connected to the fan blades through pulleys and a belt. You can change the speed of the belt driven fan by changing pulleys or by adjusting the flanges of the existing pulleys. When the belt has more than about ½ inch of play in it, you should tighten it by adjusting the motor mounts in such a way that the motor will be pushed a little farther away from the blade shaft pulley (courtesy Williamson Company.)

can save you costly replacements and repairs. It is surprising just how short a furnace fan motor will last when it is neglected. We know of several cases where neglected fan motors have burned out in less than four years. This is not to say these motors would have lasted indefinitely *with* regular cleaning and servicing, but everyone knows it is easier to prevent mechanical problems with regular maintenance, and regular inspections help you spot potential trouble spots early.

Types of Furnace Fans

There are two types of fans normally used in furnaces, and both are what you would call *squirrel cage* fans. The belt driven squirrel cage fan is shown in Fig. 4-7. This fan has a motor mounted on the outside of the fan cage, and the motor drives the fan through the use of a belt and two pulleys. If your fan has this type of drive system you will have to inspect the tightness of the fan belt periodically. Usually there will be a mechanism that is part of the motor mount that will allow you to raise the motor up a bit and away from the fan cage. This action will tighten the fan belt. At each inspection you should check the fan belt for wear and cracks that might cause the belt to break in the course of a year. There should be no more than about ½ inch of play in the fan belt when you push against it gently. If there is more, it needs to be tightened.

You also will have to be sure the belt pulleys are in line, or you may create undue stress on the belt that would cause it to break prematurely. Usually you will not have to worry about the pulleys getting out of line unless the mounting belts come

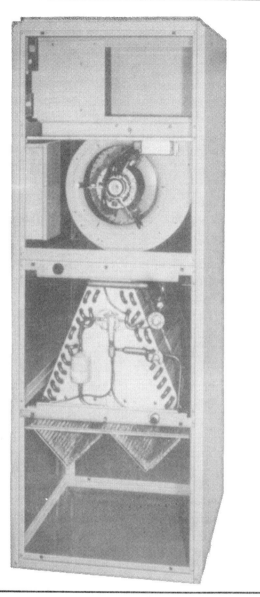

Fig. 4-8. This heat pump utilizes a direct drive fan to move air through the system. The direct drive fan has the motor mounted inside the fan cage, and the fan blades are attached to the motor shaft. One disadvantage to the direct drive fan is that when you purchase a replacement fan motor, it must be one that will fit inside the fan cage and the fan's motor mounting brackets. It does not, however, necessarily have to be the same model fan motor (courtesy Williamson Company).

loose or you remove the fan motor as described later in this section.

The second type of fan normally found in a furnace is the direct drive squirrel cage fan, shown in Fig. 4-8 as part of a heat pump installation. With this type of fan, the motor is located in the fan housing and the fan blades are attached directly to the motor shaft. There are no belts or pulleys required to connect the motor to the fan blades. You will notice that the direct drive fan requires less space for the fan assembly, which sometimes is an important factor in furnace designs.

Another type of fan you are probably familiar with that is not often used in furnaces is the bladed fan in which the fan blades are perpendicular to the fan blade shaft. This type of fan is normally used in household fans. Bladed fans are seldom used in furnaces because the squirrel cage fans move quite a bit more air for a given fan blade diameter, and will build up a higher static pressure in the duct system.

Removing the Fan and Motor

While it is not always necessary to completely remove the fan and/or motor to do a good servicing and clean up job on your furnace fan, many times it is necessary to do so. If you are faced with a problem with a defective motor, you may be able to avoid removing the entire fan assembly. On many furnaces, the motor is accessible and will come out easily by itself, but on other fan assemblies, the motor is not accessible and you will have to remove the entire assembly before you can take out the fan motor.

Removing the fan assembly varies with each furnace, so you will have to examine your own furnace to tell just how many screws or bolts must be removed and where they can be found. There will usually be a couple of screws or bolts located on each side of the fan's output flange that must be removed before you can take out the fan housing (Fig. 4-9). There will probably also be several mounting brackets that must be unbolted. Before removing the fan assembly, disconnect all attached wires and mark or tag them so you'll know where to connect them when you reinstall the fan, Fig. 4-10. If you have a downflow furnace, you may also have to remove the flue pipe or dismantle a portion of it to remove the fan assembly.

To remove the fan motor itself, you merely have to remove the mounting bolts or screws that hold it in place. This is

Fig. 4-9. This is how the fan assembly may appear to you when you remove the cover from your furnace.

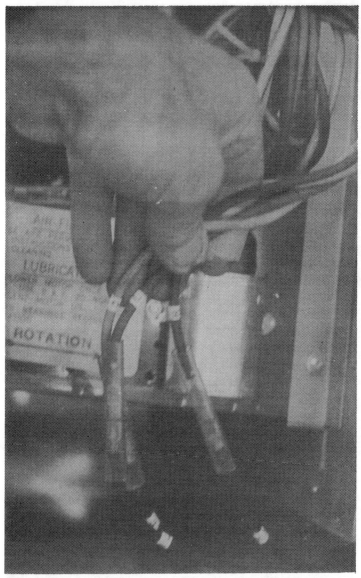

Fig. 4-10. Here the serviceman has located the fan connections and is preparing to disconnect them at these couplings. Before disconnecting the wires, however, be sure to tag them for easy identification when you reinstall the fan.

a fairly easy job once the fan assembly has been removed from the furnace, and you may also be able to do it without removing the entire fan cage. With a belt drive fan, loosen the motor mounting brackets, remove the fan belt, and the motor should

Fig. 4-11. This much dirt and grime built up on the fan blades puts extra wear on the fan motor and cuts down the amount of air the fan will move through the duct system. The blades may be cleaned with a brush, cloth or with a grease solvent. Be careful not to get any of the grease solvent in the motor! Also be careful not to disturb the fan blade weight (at pointer). This balances the blades, eases the motor load and prevents undue vibration.

come off easily. With a direct drive fan, remove the mounting bolts from the mounting flange in the fan cage. Depending on how the fan blades themselves come out, you will have to loosen the fan blade set screw on the motor shaft so the motor will come out separate from the blade assembly. Or remove the motor with the entire blade assembly attached to the motor shaft.

Cleaning the Motor and Fan Blades

As you can see from Fig. 4-11, a large amount of dirt and dust can gather on the motor housing and fan blades, even though there is a filter to remove dirt and lint from the air entering the furnace. Over a period of about a year, significant amounts of dust will build up on the fan assembly, even though you make regular filter changes. This dirt and dust can be an important energy-robbing culprit.

Dirt on the motor housing can create a sort of insulating layer on top of the motor, which can trap a good deal of the heat the motor generates. If enough dirt forms, the motor's efficiency will be cut and motor problems can set in. This surface dirt can be wiped off with a cloth, or vacuumed out with a household vacuum cleaner. Never, but never use water or a liquid cleaner to remove dirt from the motor housing unless the motor has been disassembled. The water will enter the windings and can cause a short circuit, rendering the motor useless unless the water can be dried out.

Caked-on dirt, grease and lint should be removed from the blades of the fan blade assembly. When the dirt is as thick and heavy as that shown in Fig. 4-11, the motor really has quite a bit of extra load on it. Plus, the fan blades at this point are so clogged with dirt that they are incapable of moving air through the furnace properly. A cloth or a stiff wire brush will probably remove the dirt from most fan blades. Be careful that you do not move any of the fan blade weights that may be on the blades. These keep the blade assembly in balance to ease the motor load and stop vibration. In some cases where the fan blades are really dirty, you may have to use one of the commercial grease cutting products to remove the heavy dirt. Automotive supply stores carry several spray-on products of this type. Be sure, however, not to get the liquid cleaner into the motor where it could cause a short circuit.

Fig. 4-12. This direct drive fan motor has been removed from the fan assembly and is ready for disassembly. A motor should be disassembled for a thorough cleaning and lubrication once every two years to prolong its life.

Disassembling the Motor

In the following sections, we have outlined how you can disassemble your fan motor and clean and service it upon disassembly. You probably will not have to completely disassemble your motor each year to keep it running properly. You do however, need to be sure to perform the following tasks so you won't have to dissable your motor so often:

- Keep dirt and dust off the motor housing. It is this dirt and lint that enters the motor housing and makes a periodic cleaning of the internal windings necessary. If you deep dirt and lint wiped off the outer motor housing, there won't be nearly as much to enter the motor to gum up the windings.
- Keep the bearings oiled. Oiling the motor bearings is discussed later, but you should realize that most bearings can be oiled without disassembling the motor. Usually there is an oil cup or a wick to saturate with oil accessible from outside. If you will clean the motor and inspect and oil the bearings once a year, you may be able to go as much as three years or so without ever having to disassemble the motor. As explained later in this chapter, some motors have sealed bearings, and the motor must be dissasembled to oil them.

After you have removed the motor from the fan cage mounting, place it on a workbench or a suitable area where you will have room to disassemble it. As shown in Fig. 4-12, the fan blade assembly will still be attached to the motor shaft if the fan is a direct drive fan. If it is a belt drive fan, the belt pulley will still be on the end of the motor shaft in place of the fan blades.

The blades or pulley can be removed as shown in Fig. 4-13. Loosen the hex screw that holds the hub securely onto the motor shaft. The pulley or blade assembly must be removed from the shaft to disassemble the motor, but these usually will not come off the shaft without a good deal of coaxing. You may be able to jar the pulley or fan assembly off the shaft by using a plastic mallet and a punch or metal rod placed at the base of the blades. Be careful not to damage the shaft in removing the blades or pulley. Clean the shaft with sandpaper and cleaning solvent.

Fig. 4-13. The first step in disassembling the fan motor is removing the fan blades from the motor shaft. On a belt drive fan motor, there will be a pulley in this position to remove. Loosen the set screw at the base of the blade assembly or pulley with a hex wrench. The blades should slide off the shaft, although they probably will require some coaxing in the form of tapping with a plastic mallet. Be careful not to damage the shaft! Once the blades have been removed, clean the shaft with cleaning solvent and sandpaper.

To disassemble the motor, you should first mark a line across the side of the motor so that you can line up the two bellhousings with the stator once you begin reassembling the motor.

There are several bolts that go through the entire length of the motor, from one bellhousing to the other (Fig. 4-14). These should be removed. The shaft of the motor must be clean so the shaft will slide through the bellhousing without damaging the bearing. With the bolts removed and the shaft clean, the motor is now ready to be pulled apart.

First insert a screwdriver or chisel in the crack between one of the bellhousings and the stator. A few firm raps with a hammer should be sufficient to separate the stator from the bellhousings, but you must be careful not to dent or bend the bellhousing. One bellhousing is all you need to separate from the stator, as you can see from the accompanying diagrams.

Now you can separate the motor into two halves as shown in Fig. 4-15. You should be able to get the motor to pull apart with some solid tugs on the two halves of the motor. A few careful taps with a mallet on the end of the shaft to drive it through the bearing may be needed, but be careful not to damage the shaft. And always place a small block of wood on the end of the shaft so that the mallet strikes the wood block and does not directly strike the metal shaft. This precaution will not only protect you from flying metal chips, but it also prevents damaging the end of the shaft.

Cleaning the Stator and Rotor

Figure 4-16 shows the motor disassembled and ready for a thorough internal cleaning. You should never use water or any type of cleaning solvents to clean any portion of the motor, except the rotor, as described later. Cleaning solvents may dissolve the protective, insulating coating that covers the wires of the windings, and this will cause a short circuit that will ruin the motor. Water can cause rust that is equally damaging to the motor. Scraping the windings is also forbidden for the same reasons.

To clean the stator, first wipe the windings clean with a soft cloth. Compressed air can also be used to blow out some of the dust and dirt inside the windings; but do not get the nozzle too close to the windings with high air pressure. Any

Fig. 4-14. Before you begin separating the motor parts, mark a line across one side of the motor from one bellhousing to the other. This line should be marked with a scratch mark or non-smearing ink. This will help you line up the motor parts correctly when you reassemble the motor. Two are indicated at (A). By inserting a screwdriver in the crack between the front bellhousing and the stator at (B), you can separate the motor into two halves.

remaining dirt should be wiped out with a cloth. Again, remember to never wash or scrape the windings! The stator and bellhousing ready for cleaning are shown in Fig. 4-17.

The rotor, shown in Fig. 4-18, is attached to the motor shaft and should remain attached throughout the cleaning. The rotor and shaft should be removed from the front bellhousing and bearing. Since the rotor contains no wires that might be damaged from cleaning solvent, it is okay to wipe the rotor clean with a cloth and cleaning solvent.

The motor shaft will be completely exposed for cleaning at this point. The shaft should be free of rust and dirt, and may be wiped clean with a solvent. You also may want to sand the shaft smooth as necessary to remove rust and to allow the bearings to slide back onto the shaft easier. But do not sand those portions of the shaft where the bearings will be because sanding at these points can create bearing wear. The shaft must be polished at these points.

Bearing Lubrication

Making sure the bearings are lubricated is one of the most important things you can do to insure prolonged motor life. It is usually not necessary to disassemble the motor to lubricate the bearings (see Fig. 4-19), but disassemby is necessary to really clean thoroughly and relubricate them. The bearings are located in each bellhousing where the shaft passes through the housing. They should be removed and cleaned with a grease solvent to remove as much dirt, grime and old grease as possible. Then, lubricate them generously with SAE 10 motor oil. If there is a cloth wick that is part of the bearing housing, saturate it with oil also, to keep the bearing lubricated long after it is reinstalled. Move the bearings around to be sure they work properly. If there is any grittiness or drag in their movements, you should replace the bearings unless a second soaking in grease solvent will remove grit that may not have come out the first time. Replacement bearings can be purchased at any supply house dealing with electrical motor parts.

The usage of what has come to be called *lifetimes* bearings has led many homeowners to think that their "lifetime" bearings need no maintenance. How wrong this assumption is! Although it is true that a "lifetime" or sealed bearing can go

Fig. 4-15. After the through bolts have been removed, the motor is separated into two halves for cleaning and lubrication. Only one of the bellhousings needs to be separated from the stator. Here, rotor (A) and front bellhousing (B) are separated from stator (C) and rear bellhousing (D).

longer than a conventional bearing without relubrication, it will eventually need servicing, also. Usually you will find these bearings installed on newer electric motors.

These bearings are porous and have tiny holes in them that force the oil to pass through the bearing. They have a reservoir to hold a supply of oil. You can add oil to these bearings by saturating the bearing with oil.

What we have described so far is a regular maintenance program for your motor bearings. An external motor cleaning and bearing lubrication once a year will usually be sufficient, and once every two years, you should disassemble the motor to thoroughly clean the bearings and relubricate them. Of course, if a bearing goes out so that it freezes, sticks or begins squeaking, immediate attention is required. Disassemble the motor, remove the problem bearing, and lubricate it if possible. You may find that you'll have to replace the bearing.

Reassembling the Motor

To put the motor back together, you simply repeat the steps you used in disassembling it, but this time, of course, the order is reversed. First, place the rotor in the front bellhousing. Next, place the rotor with front bellhousing into the stator-rear bellhousing portion. Push the two motor halves together, and line up the marks you made earlier across the side of the bellhousings and stator. Turn the motor shaft to be sure it rotates freely, insert the through bolts, and reattach the fan blades or pulley.

ADJUSTING THE FAN CONTROL

The adjustment of the fan control plays an important part in the operating efficiency of your furnace because the fan is the mechanical device that moves heat from the furnace into the living area. If the fan turns on too soon after the furnace begins heating, it will blow cool air into the living area and create a cold-air draft. If it turns off too soon, there will be a lot of heat remaining in the furnace that will be wasted up the flue and will never be blown into the living area.

Furnace fan switches are separate from the thermostat, so the fan will turn on only when sufficient heat has been

Fig. 4-16. Now the two motor halves have been separated and are ready for cleaning. You may have to tap the end of the motor shaft with a mallet to drive the shaft through the rear bellhousing bearing to separate the motor. When doing this, be careful not to damage the end of the shaft, and always place a block of wood against the shaft end so that the mallet strikes the wood block and not the shaft end itself.

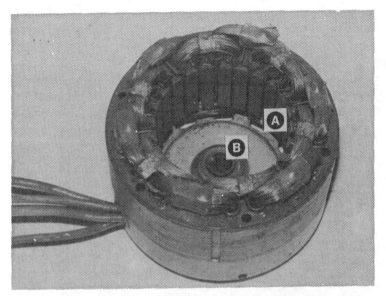

Fig. 4-17. The stator and rear bellhousing remain attached for cleaning. You should wipe the stator winding (A) clean with a clean cloth. Do not scrape the windings, use water or any sort of solvent for cleaning, as this can damage the protective covering of the wires in the windings and cause a short circuit. The bearing (B) should be removed for soaking in a grease-cutting solvent. The bearing should be thoroughly flushed to remove any grit or rust inside it. After allowing the solvent to dry out of the bearing, work it back and forth to be sure it moves freely, and replace it if it does not.

generated in the heat chamber. Usually they are heat sensitive bimetal switches located at the heat chamber to sense the temperature there. Once the temperature reaches the fan turn-on temperature of, say, 150°, the switch closes and the furnace fan begins blowing heat into the living area. But this happens *only* after the furance burners have had time to generate some heat in the furnace. If the fan came on with the burners as soon as the thermostat called for heat, the fan would do nothing more than push a lot of cold air through the duct system and create a cold draft, until the burners warmed up the heat chamber.

Likewise, the fan does not turn off when the thermostat cuts off the burners. When the burners stop heating, there is still quite a bit of heated air left in the furnace. So the fan stays on long enough to blow this heated air into the living area. Of course, with no heat being generated by the burners, it isn't long before the temperature in the heat chamber is cool once again. When the temperature cools to, say, 125° in the heat

Fig. 4-18. The Rotor (A) should remain attached to the motor shaft throughout the cleaning. The rotor and shaft, however, must be removed from the front bellhousing and bearing so the rotor, shaft and bearing (B) can be cleaned. The rotor itself may be cleaned with a solvent, since there are no wires or windings in it that can be damaged. Sandpaper should never be used to clean the rotor. Clean the motor shaft with solvent and sandpaper, but be sure not to sand the portion of the shaft that touches the motor bearings. Doing so can escalate the bearing wear, so the shaft should remain polished at those points. The bearing should be cleaned and oiled.

chamber, the fan control turns off the fan. This is the fan turn-off temperature.

The fan switch on most gas furnaces (Fig. 4-20) is adjustable, and the fan switch on an oil furnace is similar. By moving the levers on the switch, you can adjust the turn-on and turn-off temperatures of the furnace fan higher or lower as needed. An electric furnace fan switch is usually not adjustable.

A furnace has a limit switch to keep the furnace from overheating should the fan not turn on, and this switch is frequently a part of the fan control switch. The limit switch is a protective device that turns off the burners if the fan doesn't blow the hot air out of the heat chamber. Without it, the burners would continue heating and building higher and higher temperatures until the furnace were damaged or a fire began. The limit switch is set by the factory for each furnace, and it should not be adjusted. In an electric furnace, the limit switch takes the form of a bimetal switch or link that turns off the furnace power when temperatures in the heating element bank gets too high.

The factory setting on the limit switch is usually less than 225°. In Fig. 4-20, the limit switch is set at 170°. You will do nothing to improve furnace performance by tampering with this switch. Raising the limit temperature makes the furnace unsafe, and lowering it may reduce furnace efficiency. While the limit switch may be part of the fan control switch, as shown in this diagram, it may also be a separate switch in some furnaces.

As already mentioned, the fan controls on electric furnaces are normally nonadjustable controls installed by the factory for the most efficient furnace operation. The only way to change the turn-on and turn-off temperatures of such controls is to purchase new controls, but this is generally not advisable unless you are having significant problems, such as *short cycling*, that you suspect are due to the fan controls.

Short cycling is a defect in fan control settings that usually appears in the adjustable fan controls—and then only when the controls have been set improperly. Short cycling occurs when the furnace heats up, the fan begins running, and then the burners stop heating abruptly so that the furnace cycle is perhaps less than a minute long. Then the furnace will turn on again shortly, and the process is repeated.

Fig. 4-19. It usually is possible to oil the bearings on electric motors without dismantling the motor. Although this motor has been disassembled for cleaning and lubrication, it is possible to lubricate the bearings by placing a few drops of oil into the hole in the bellhousing indicated by the pointer. Motors with sealed or "lifetime" bearings will not have an external oil hole. On some motors you will find a cloth wick located in the bearing housing near the shaft. This wick can be saturated with oil to lubricate the bearing.

Short cycling, as you can see from the description, robs the furnace of energy because the furnace spends so much time turning on and off. It never reaches a temperature that is really warm enough to heat the house. There are a number of possible causes of short cycling, including a stopped-up furnace filter or a blocked air flow of some other type. If air cannot flow through the system, heat cannot escape from the heat chamber and the limit switch will cut the furnace off, causing short cycling. Another cause of short cycling is an improperly located thermostat. If the thermostat is located right by a heat register, once the heat blows through the register, hot air of temperatures as high as 125° will come through the register. If the thermostat is directly in line with that hot air, it will sense the very warm temperatures and cut the furnace off, only to turn it on again when the hot air has dispersed.

Short cycling can also be caused by improperly set fan controls, as already mentioned. This occurs when there is too little difference between the fan turn-on temperature (the higher temperature) and the fan turn-off temperature (the lower temperature). The fan control will usually have two levers that can be adjusted to change these temperatures. If your control has three levers, one will be the limit switch control and should not be adjusted.

Normally you will not have short cycling problems if you set the fan controls so there is at least a 25° temperature difference between the fan turn-on temperature and the fan turn-off temperature. Thus, one common setting is to have the fan turn on at 150°, and to turn off at 125°. If less temperature difference is set, however, short cycling problems can happen easily. Suppose you set the fan control for a turn-on temperature of 150° and a turn-off temperature of 140°. When the fan begins heating, the fan will not turn on until the heat exchanger temperature reaches 150°. But as soon as the fan starts, it will draw quite a bit of 70° air into the furnace from the return air duct. Naturally, that cools off the heat exchanger quickly, and soon the temperature has dropped below the 140° turn-off temperature, even though the burners are going. The fan control shuts off the fan. But soon the heat exchanger is reheated, and the fan starts up once again, repeating the cycle. These problems are avoided if about 25° in temperature difference on the turn-on and turn-off settings is maintained.

Fig. 4-20. This is a combination fan control and limit switch that is similar to those found on many gas and oil furnaces. The fan controls of this switch are adjustable, but the limit control is not. To change the fan turn-off temperature, move lever (A) at the side of the dial. To change the fan turn-on temperature, move lever (B). Slot (C) is cut into the dial to set the limit control permanently at 170°.

You can make your furnace operate more efficiently by lowering the turn-on and turn-off temperatures below the common 150° and 125° settings mentioned earlier. If the turn-on temperature is lower, the burners do not have to get the heat chamber parts so hot before the fan begins blowing warm air out. Thus, the burners do not expend so much energy warming the heat exchanger walls and the rest of the heat chamber all the way up to 150°, only to have it cool off when the burners shut down. A turn-on temperature of about 120° to 125° will turn the fan on soon after the burners begin, and the furnace air should still be plenty warm, although you may notice it will be a bit cooler than you might be used to.

Lowering the furnace fan turnoff temperature down to 90° to 100° is another efficiency increaser because the fan runs long enough to blow almost all the generated heat out of the heat chamber. If the fan stops at 125°, it is apparent that there would be a lot of heat left in the furnace system that is simply lost. Setting the turn-off temperature below 90° is not likely to do much good because at that temperature the air in the furnace is really little warmer than room temperature, and the air blowing out of the heat registers feels cool. Figures 4-21 and 4-22 show how to locate the fan control on your furnace for adjustment.

Settings at 90° and 125° for the fan switch may be too low for your home. You need to take some time to adjust the settings, wait for about a day, and then readjust the fan turn-on and turn-off temperatures. Setting the fan turn-off temperature as low as possible will save the most energy.

MORE ON FURNACE TUNEUPS

This chapter has dealt primarily with tuneup procedures that are common to all warm-air central heating systems: gas, electric and oil furnaces, and electric heat pumps. Most of the tune up procedures included in this chapter are also applicable to most modern coal and wood furnaces. This chapter is by no means an exhausive list of tuneup procedures for *any* heating system. The following chapter deals with specific tune up procedures for gas, oil and electric furnaces. Wood heat system maintenance is discussed in Chapter 9 (wood heat), and heat pumps are discussed in Chapter 10 (heat pumps). So in short, the tune up procedures outlined in this chapter are

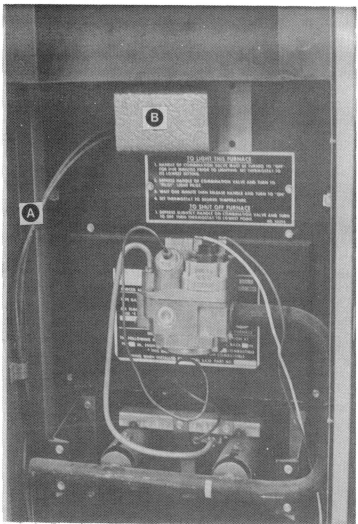

Fig. 4-21. You can locate the fan control on your furnace by tracing the wires (A) connected to the fan motor until they reach a control box (B).

simply a starting point. For purposes of organization, we have placed tune up procedures for specific heating systems in the chapters dealing with those heating systems, and tune up procedures for specific furnace parts are also in chapters that discuss those parts.

Beyond the procedures we have outlined in this chapter, your once-a-year heating system tune up should include:

- Any specific maintenance procedures for your particular heating system. Most of these are outlined later in this book.
- Cleaning, scraping, and servicing the humidifier in your system. This is discussed in Chapter 12 (Humidifiers).
- Cleaning and inspecting your thermostat, as discussed in Chapter 11.
- Cleaning your furnace, checking wiring to be sure wires are securely connected, and a visual inspection to spot any potential trouble spots that may require additional attention.

ENERGY EFFICIENCY CHECKLIST

As part of any initial furnace tune up, you should:

- Clean the furnace filter and inspect it for defects if it is a washable or an electronic air filter. Replace the furnace filter with a new filter the same size if it is a disposable air filter. Be sure to check to see if your furnace has more than one filter.
- Remove the fan assembly, clean the fan, disassemble the motor for cleaning, clean and relubricate the motor bearings.
- Examine the fan control switch. If the switch is adjustable, be sure there is at least a 25° temperature difference between the turn-on and turn-off temperatures. For most efficient operation, set the turn-on temperature for 125°, and the turn-off temperature for 100° or less. If these temperatures turn out to be unacceptable, you may have to readjust these temperatures after the fan has been in operation a while.

After the initial furnace tune-up, follow this timetable:

Filters. Every air filter should be cleaned (if electronic or washable type) or replaced (if disposable type) at least once each year at the beginning of each heating season. Inspect washable filter elements for tears that indicate replacement is needed. About the middle of the heating season, clean washable filters, and replace disposable filters, unless inspection shows disposable filters do not need replacement. Vacuuming and lightly cleaning disposable filters every month or so will prolong their life and increase furnace air flow.

Fig. 4-22. Removing the cover of the control box reveals the fan control switch underneath.

Fan and motor. At beginning of the heating season each year, wipe fan motor and blades clean with a brush and cloth, inspect fan motor mounts, inspect belt for wear, tighten the belt if necessary, and oil bearings. Once every two years disassemble the motor for complete cleaning and lubrication.

Fan Controls. Once the controls are set, they should not require readjustment.

Tuneup And Servicing Techniques For Oil, Gas And Electric Furnaces And Boilers

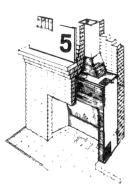

In Chapter 4 we outlined some of the things you can do to tuneup almost every furnace, no matter what the fuel. Of course, cleaning filters and fans, important as it is, is only a starting point. Each type of furnace has a different combustion system and different controls that must be in adjustment for you to get the maximum out of your furnace. In addition, each type of furnace presents a different set of repair of problems for the homeowner and serviceman. In this chapter we deal with tuneup procedures and common servicing problems for oil, gas and electric furnaces, as well as boilers.

Although the illustrations presented in this chapter and much of the discussion are devoted to oil, gas and electric furnaces, virtually the same techniques are used to service hot water heating systems. Oil boilers and gas boilers, especially, have combustion systems very close to their warm-air furnace counterparts. Thus, if you have a boiler, the service and tuneup techniques in this chapter are applicable to your heating system, as well.

The illustrations and diagrams we have used in this chapter are, by necessity, general in nature. Specific illustrations for each type of furnace or boiler available would simply be too voluminous. Therefore, we have included pictures and diagrams that will give you a general idea of tuneup and servicing principles. The basic parts for each type of heating system are pretty much the same, but you will have to use your own ingenuity and the owner's manual or service manual for each particular unit to get specific instructions on access to each furnace part. What we have done in this chapter is to outline for you specific steps you should take in tuning up or servicing a heating system. Locating the specific parts to be adjusted for any particular unit will have to be left to you. But if you will study the illustrations here you will know what you are looking for, and you will have a good idea of how to reach them. Your own skill and ingenuity will make every furnace part accessible.

OIL FURNACES

Oil furnaces have been increasing in popularity lately with new home builders in many regions where oil has not been traditionally a popular fuel. One of the largest reasons for this is the growing problem of securing natural gas supplies for home installations. In many regions where natural gas shortages have become chronic, no new residential hookups onto natural gas lines are allowed. In search of an inexpensive heating fuel substitute, many homeowners have turned to oil. In most areas, oil is considerably less expensive than electric resistance heat, and is cost competitive with electric heat pumps.

OIL FURNACE OPERATION

Chapter 2 (What You Should Know About Your Heating System) introduces you to the basics of oil furnace operation. The most popular residential oil furnace is the gun type oil burner, in which oil is drawn from a supply tank into a pump that is part of the burner assembly. This pump places the oil under high pressure and sprays it through the oil burner's nozzle. Here, the oil mixes with air and comes in contact with a high-voltage arc across the electrodes at the end of the nozzle. This electrical arc ignites the fuel, much as a spark plug ignites the fuel in a car, and the fuel burns inside the combustion chamber. The flame warms the walls of the heat exchanger, and air blowing over the outer wall of the heat exchanger carries the heat into the living area. Figure 5-1 shows the parts of the oil burning furnace in operation.

Safety Controls

The oil furnace has two safety controls. A stack control safety switch located in the furnace flue senses the heat going up the flue, and as long as the stack control probe is being

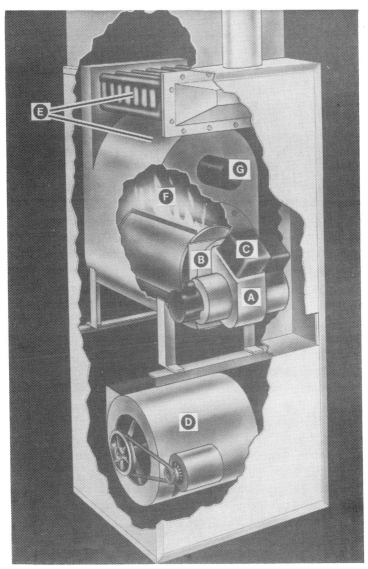

Fig. 5-1. An oil-fired furnace (A) Gun-type burner assembly (B) Burner mounting flange. (C) Burner control box. (D) Squirrel cage fan, belt driven. (E) Heat exchanger. (F) Combustion chamber. (G) Flue. Courtesy York Division of Borg-Warner Corporation.

warmed, it will allow the pump to continue operating. But if for some reason the furnace's igniters are defective and do not light the oil, the probe will not be warmed and the safety switch will turn the pump off. This will prevent the pump from spraying unlighted oil into the combustion chamber and causing a fire hazard.

When the thermostat calls for heat, the stack relay turns the igniters on and starts the pump. Depending on the burner's design, the igniters may stay on after the oil lights, or they may turn off. Heat that escapes up the flue heats the probe on the stack switch, and the stack switch holds the pump relay on. But if the stack control probe is not warmed in the proper amount of time, the control will turn the pump relay off and the pump quits.

If the stack control shuts off the furnace, you will have to push a button on the face of the control to get the stack control to recycle. This recycling will remove any unburned oil from the firebox and set the unit up to start again. The burner should go through its starting cycle once more, and will probably light this time unless there are ignition problems. Sometimes the stack control will turn off the pump because of a cold gust blowing down the flue momentarily. If you recycle the burner, usually it will run. If you have ignition problems or the stack controls turns the burner off a second time, something else is wrong and servicing is called for. This is discussed later.

The second safety control in the oil furnace is the *limit switch*, which was discussed in Chapter 4 in the section, Adjusting the Fan Control. This switch is located near the heat exchanger and usually has a heat-sensitive probe that extends into the heat exchanger area. If the fan does not turn on for some reason, the limit switch will sense the high temperatures building up in the heat exchanger and will turn the burner off to prevent furnace damage or a fire. On most oil burning furnaces the limit switch will be part of the fan control switch, as described in Chapter 4.

Electrical System

The oil burner has two transformers. One steps down the voltage from 120 volt line voltage to 24 volts, which is connected to the control circuit with the transformer, relays and stack control. Figure 5-2 shows the wiring diagram for the

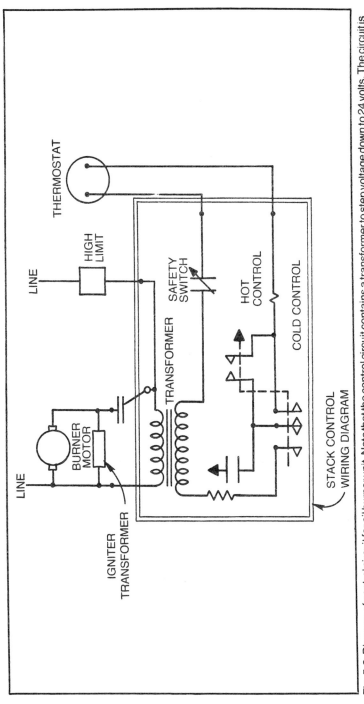

Fig. 5-2. Diagram of control circuit for oil burner unit. Note that the control circuit contains a transformer to step voltage down to 24 volts. The circuit is activated by the thermostat. The area inside the box is the stack control, or the primary control circuit.

control circuit. This control is the stack control, previously described, and it includes the motor and ignition relays, as well as the stack control safety switch. When the thermostat calls for heat, these relays are energized and the pump starts and the igniters arc. This stack control will normally be mounted in the flue, where the safety switch probe can sense the heat going up the flue.

The second transformer steps up the voltage from 120 volts to 10,000 to 15,000 volts used to arc at the electrodes and light the fuel. The electrodes are made of stainless steel in a ceramic insulator, and they must be positioned and gapped correctly for the oil burner to work properly. Setting the electrode gap is discussed momentarily.

FUEL TANK AND FILTER

Fuel oil for an oil burner must be stored in a nearby storage tank, and this tank must be refilled periodically by the oil supply company. The tank should not be located too near the furnace, however. For safety, always have at least 7 feet between the burner and the storage tank.

The oil may be stored inside or outside, but outside storage can cause problems in extremely cold weather. Sub-zero temperatures can turn the fluid oil into a slow-moving sludge that will not flow through the transfer lines. If you have been having cold weather problems with your oil burner, you should consider moving the tank indoors, if possible. If your oil burner is located in a basement, a favorite storage tank location is in the basement with the burner. Remember to keep the 7 foot distance!

The tank is connected to the oil burner through two different fuel line systems. In a two-line system, part of the oil delivered to the burner is returned to the tank. In a single-line system, all the oil delivered to the burner is burned.

Somewhere in the fuel line between the storage tank and the burner, there should be a filter to trap dirt and moisture that might otherwise enter the burner pump and nozzle. See Fig. 5-3. This filter should be cleaned as part of the routine oil burner tune up and maintenance, described in the following section.

OIL BURNER TUNEUPS

Whether you own an oil furnace or an oil-fired hydronic system, tuning up your combustion system can save you a lot

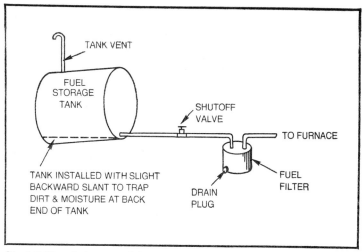

Fig. 5-3. The oil storage tank has a shutoff valve and a filter in the line between the tank and the burner. This filter must be drained and cleaned as part of the routine maintenance. The storage tank should be installed with a slight backward slant. This allows dirt and moisture to settle in the tank where it will not enter the fuel system. Of course, occasionally this tank must be drained to remove the accumulated dirt and moisture.

of money. Most oil-burning systems are capable of running at 75 to 80 percent efficiency, but they must be in good tune. (This figure would not account for later heat losses in the duct system, which would make the overall heating system efficiency somewhat lower). But if your oil burner has not been serviced and tuned in several years, it may be operating at only 60 percent efficiency, or even lower. This represents a tremendous waste in fuel costs each day you have your heating system turned on. Increasing your heating efficiency from 60 percent to 80 percent will save you a whopping $33 for *each* $100 in your annual fuel bill.

Most portions of an annual oil burner tuneup can be performed by the homeowner with a minimum of tools. In some cases, it might be advisable to contact a professional oil burner serviceman to perform a complete efficiency test on your oil burner. By performing this test, the serviceman can change the burner nozzles to reduce oil consumption to the optimum level for your installation. This can be a one-time-only adjustment that will keep your oil burner saving energy for years to come. The professional adjustments are discussed later in this chapter.

Whether or not you have any professional adjustments made on your oil burner, you should perform a complete tuneup yourself on the unit once a year before the start of the heating season. Then, if you do decide to pay for professional oil consumption adjustments, you will only have to pay the serviceman for those adjustments and not for a complete tuneup.

Visual Inspection and Cleaning

A large part of the tuneup of any combustion system involves a careful visual inspection of all working parts, connections and wiring terminals to spot any potential problems. Check wiring connections and terminals to be sure they are sound and tight. All tubing connections should be tight so they will not cause oil leaks. Since a tuneup involves dismanteling a good deal of the combustion system, you should wait until the end of the tuneup to secure wiring and tubing connections.

Air leaks in the combustion chamber and heat exchanger can cause a variety of problems, including burner inefficiency, soot buildup, and smoky furnace air. A removable panel or door should give you access to your heat exchanger and combustion chamber. If there is heavy soot buildup on the inside of the heat exchanger walls, it should be brushed away and vacuumed off. This soot acts as an insulator to trap heat inside the heat exchanger so it can't reach the air blown around the heat exchanger walls.

If substantial soot buildups are present on the heat exchanger walls, the flue also needs cleaning. Soot buildup there can cause the stack control to malfunction.

Inspect the combustion chamber (located below the heat exchanger where the burner nozzle extends and the fire burns). Large soot buildups inside the combustion chamber are also insulators that trap heat. There should be no air leaks in the combustion chamber.

Cleaning Fuel Filter

The fuel filter removes harmful dirt and moisture from the fuel traveling from the storage tank to the burner. Naturally, if this filter is to retain its effectiveness, it must be removed and cleaned. Usually this filter will be a sediment bowl type filter that is simply drained for cleaning. Sometimes, however, a disposable paper filter is included in this filter. The disposable element should be thrown away, but be sure you have a replacement before you dispose of it! Replacements should be available from your fuel supplier. Once you locate replacement filter elements, purchase at least one extra. That way, if your filter ever gets clogged to the point that replacement is mandatory, you'll have a spare handy to replace it.

To clean the fuel filter, you should first close the storage tank shut-off valve. This will usually be located in the fuel line next to the tank. The filter is sometimes connected to this valve, but in some cases it may be found closer to the burner.

Drain the filter bowl by opening the drain plug on the bottom of the filter bowl. After the filter is drained, remove the bowl and wipe it clean inside with a clean cloth.

While you are cleaning the filter, give the supply tank a visual inspection. Look especially for evidence of oil dripping onto the floor. Such oil forms a dangerous situation and should be removed as soon as possible. Tighten all connections to stop the oil spillage.

Fig. 5-4. This is the front view of a pedestal-mounted gun-type oil burner that has been removed for cleaning. (A) is the barrel that houses the nozzle assembly and electrodes. When installed, the barrel extends into the combustion chamber. (B) is the motor that powers the unit. (C) is the pump that draws fuel from the tank and forces it through the nozzles under high pressure to form a fine mist. (D) is the fuel line from the pump to the nozzles.

Cleaning and Servicing the Burners

The burner is the heart of any oil-fired furnace or boiler. It is where the fuel ignites and heat is generated. At the burner you will find the pump, nozzle, electrodes and air adjustment, all which need servicing as part of your routine maintenance program. The gun-type oil burner is by far the most common oil burner type, and it is the only type we will deal with here.

You can service the oil burner without removing it from the furnace, but you may have to remove it to clean the combustion chamber. You will find the burner is mounted on a pedestal at the burner's base or it will be flange mounted onto the side of the furnace. Figures 5-4 and 5-5 show the parts of one oil burner model, which happens to be a pedestal-mounted type. Also see Fig. 2-13. Figures 5-1, 5-6 and 5-7 show a flange-mounted burner.

Before removing the burner from the unit, first double-check to be sure the power is turned off. Then disconnect the power supply from the burner, and disconnect the fuel lines from the burner. Of course, you should also double-check the fuel lines to be sure they are closed.

If you look into the open end of the burner's barrel, you will see the nozzle and the electrodes (Fig. 5-8). Notice that the electrodes are located above the nozzle and not directly in front of it. The electrodes and nozzle assembly can be removed for cleaning and adjustment by removing the plate at the back of the barrel (Fig. 5-9). There should be at least one screw or bolt anchoring the nozzle assembly to the barrel. Remove the anchoring screws and the wires connecting the igniter to the transformer. The igniter assembly will now slide out the back of the barrel and is ready for cleaning (Fig. 5-10).

Remove the nozzle and place it in a cleaning solution. If the nozzle appears cracked, replace it. Inside the nozzle at the fuel connection will be a small screen filter that also should be cleaned.

Scrape the electrodes clean and sand them. If they are broken, or if part of their ends have burned off, replace them. Check the porcelain bases for cracks. These insulators must be kept clean.

There will be a bolt or screw clamp holding the electrodes in place on the igniter. If electrode adjustment is needed, they should be adjusted according to the manufacturer's recommendations. If you do not have an owner's manual or a service

Fig. 5-5. Rear view of the same burner. This is the view you would see looking at your furnace when the burner is installed. (A) motor (B) pump. (C) Transformer that steps the line voltage up from 120V to more than 10,000V for the igniters. (D) Removable plate under which the igniters are found.

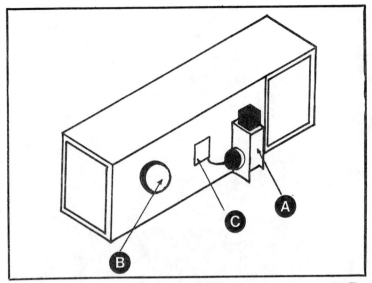

Fig. 5-6. This horizontal oil furnace uses a flange-mounted oil burner (A). The burner is attached to a flange that bolts into the furnace cabinet. B is the flue. C is the stack control where relays and switches are located.

manual, follow the specifications shown in Fig. 5-11. The electrodes are positioned in front and above the nozzle. The gap between electrodes should be ⅛ inch to 3/16 inch. The electrode tips should be ½ to 5/8 inch above the nozzle, as measured from the center of the nozzle. The position of the electrodes in front of the nozzle depends on the nozzle spray pattern. No part of the electrodes would be less than ¼ inch away from metal parts.

Pump Filter

Inside the pump is a circular screen filter that keeps dirt and sludge from entering the pump and fouling up the burner's fuel system. Over a period of time, this filter will become so filled with sludge that it may nearly stop the flow of oil through it.

To get to this filter, remove the five to eight hold-down bolts on the outer pump casing that holds the outer plate onto the pump (Fig. 5-12). Underneath this plate, there will be a filter lining the walls of the pump. In Fig. 5-13, this filter has been removed and has been placed on top of the pump.

The filter screen is delicate, so do not scrape it or poke it with a sharp object. Instead, soak the filter in kerosene or clean fuel oil, and remove the sludge with a clean cloth and brush.

Air Flow and Pump Pressure

The final part of your annual tuneup is to check out the oil burner's actual operation to see if it is operating properly. For this you will have to reinstall the burner into the combustion chamber (if it was removed) reinstall all parts and reconnect the fuel lines and electrical connections. You will have to prime the pump by filling it with fuel oil before starting. Normally there is a cap nut or a plug at the top of the pump housing for this purpose. If you have a gravity flow oil line, bleed the air out of the line before starting the burner.

You can check the pump pressure with a pressure gauge installed onto the pressure fitting on the pump (Fig. 5-14).

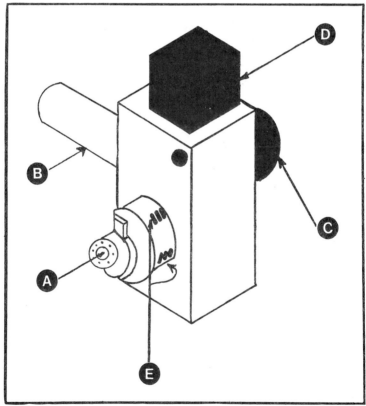

Fig. 5-7. Flange-mounted oil burner. A is the pump. B is the barrel. C is the motor. D represents the igniter transformer. E is the air adjustment.

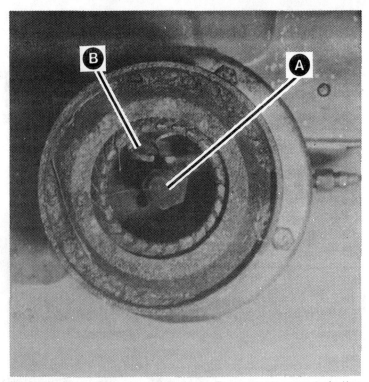

Fig. 5-8. This is the view looking into the barrel. The nozzle (A) sprays the fuel in a fine, atomized mist. The mist is ignited by electrodes (B). Depending on the burner design, the electrodes will continue arcing as long as the pump sprays oil. On some models electrodes will stop arcing and allow the fuel mixture to continue to burn on its own after it is lit.

Follow the manufacturer's instructions in setting the pressure. Usually the pressure will be about 100 psi. The pressure can be adjusted by turning a screw or nut. Sometimes this will be located under a cap nut. If the pump will not reach the specified pressure, it may be because you have not installed the removable plate tight enough, the filter screens may be clogged, or it may be that the pump parts are worn. If wear is the problem, you will have to replace the pump or the parts.

After setting the pump pressure, you can adjust the air flow to the burners. Although the best way to set the air flow exactly right is with one of the analyzing tools discussed in the next section, you can set the air flow reasonably close by visual inspection. You can usually see the flame by looking through a peep hole somewhere on the burner or furnace assembly. Sometimes it is necessary to use a special small mirror on a long handle devised for oil burner flame inspection.

The proper flame should burn yellow, making just a bit of smoke. Also, the electrode tips should not touch the oil spray. If there are smoky tips to the flame or the flame is reddish or orange-colored, it means there is too little air reaching the flame. On the other hand, allowing too much air to reach the flame leads to inefficient burning. The air adjustment is normally located near the pump, on the outside of the burner housing. Once you have set the air to the right level, tighten the adjustment mechanism with the screw provided.

Professional Tuneups

Thanks to increased interest in energy conservation, spurred by dwindling oil supplies and rising fuel costs, quite a

Fig. 5-9. The igniter assembly is located under the removable plate at the back of the burner. To remove the assembly for cleaning and adjustment, you will have to disconnect fuel line (A) and transformer lines (B) Anchoring screws or bolts will also have to be removed. Then the igniters should slide out backwards.

Fig. 5-10. This is the igniter assembly removed and ready for cleaning. You should scrape the accumulated carbon from the nozzle and electrodes. The nozzle should be removed from the assembly for complete cleaning, and the electrodes should be gapped. If either the electrodes or nozzle indicate they are broken or defective, replace them.

bit of research has been done recently on oil burner energy conservation techniques. Or perhaps it is more accurate to say that quite a bit more *interest* has been generated for techniques that have been available for years.

Government reports indicate that many oil burners across the United States are oversized for the heat loads they have to handle. This is especially true in cases where an oil burner was installed many years ago in an uninsulated home that since has been fitted with insulation. Oversizing means inefficient burning and wasted heating fuel dollars.

A relatively simple way to correct oversizing with oil burners is to install a smaller capacity nozzle on the burner so not as much fuel is burned. But this is a job that is not really tailored for do-it-yourselfer.

To do the job right and get the optimum efficiency out of the oil burner, the serviceman must perform smoke tests, carbon dioxide test, and flue gas temperature measurements on the unit. By performing these tests and comparing the results with a scale, he can select exactly the right nozzle for your oil burner. But the testing equipment alone to perform these tests costs well over $100, which is out of the reach of the do-it-yourself homeowner.

We offer this recommendation. Paying a professional serviceman to select the proper nozzle for your oil burner can be money well spent, because it will increase the efficiency of your unit for years to come. And it is a one-time purchase of his services, so you don't have to pay him to come back each year. Once you have the best nozzle, that's it! In subsequent

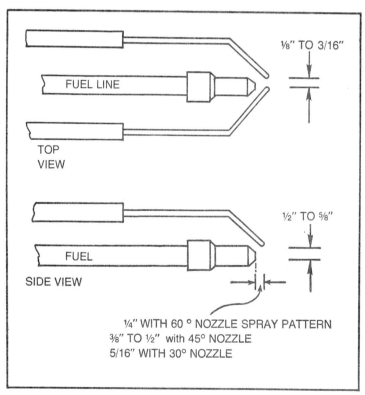

Fig. 5-11. If you have access to your oil burner's owner's manual or service manual, get the electrode specifications there. If not, these specifications will work in most instances. Note that the distance the electrodes are placed in front of the nozzle depends on the spray pattern of the nozzle.

Fig. 5-12. This is the removable plate at the side of the burner pump. By removing the hold-down bolts, you can remove this plate and clean the pump filter.

years you can do the tuneup yourself, following the procedures outlined in this chapter.

If you do decide to hire a serviceman to perform all the tests required to select the proper nozzle for you, follow the tuneup procedures already outlined before he arrives so the bulk of the tuneup work will be done. Before the tests are performed, a routine tuneup must be given to the oil burner. If you do it before he arrives, you save the cost of his doing it. When you ask the company to send a serviceman, tell them you want him to perform a carbon dioxide measurement, smoke test and flue gas temperature measurement. Also, you will want him to install the optimum nozzle size on your oil burner. If the company you contact cannot do this, get another company.

If you are a part-time or full-time serviceman interested in learning how to optimize nozzle sizes, you should check with your heating supply house for an oil burner efficiency testing kit. You should also obtain the publication "A Service Manager's Guide to Saving Energy in Residential Oil Burners: Optimizing Nozzle Size" from the Department of Energy, Washington, D.C. If you perform a lot of oil burner tuneups you really should purchase one of these efficency testing kits because they will help you set the air adjustment exactly right.

OIL BURNER TUNEUP CHECKLIST

1. Visual inspection and cleaning. Remove soot buildup in heat exchanger and flue.
2. Clean fuel supply line filter and replace disposable filter element.
3. Clean nozzle and nozzle filter.
4. Scrape electrodes and insulators.
5. Set electrode gaps.
6. Clean pump filter.
7. Check and adjust pump pressure.
8. Adjust air flow until flame is a yellow color.
9. Adjust flame professionally with carbon dioxide

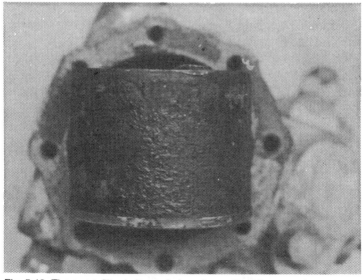

Fig. 5-13. The pump filter is located along the walls of the pump beneath the removable plate. This filter has been removed and placed on the pump in preparation for cleaning. As you can see, the filter is filled with sludge that makes oil flow quite difficult. Clean the filter in kerosene or fuel oil.

analyzer, smoke tester and flue gas temperature tester. Optimize nozzle size.
10. Tighten all electrical connections and fuel lines.

FURTHER OIL BURNER SERVICING AND TROUBLESHOOTING

A once-a-year tuneup and inspection will go a long way to making an oil burner trouble free. This section is devoted to other common oil burner servicing problems and techniques.

Igniters

Sometimes "backfire" problems will develop in an oil burner. This happens when ignition is delayed, giving the pump time to spray a large amount of oil mist into the combustion chamber. When the oil is ignited, a small explosion occurs, sometimes sending soot and oil back into the furnace room. The problem may be caused by insufficient voltage across the electrodes. Check the line voltage and the voltage from the transformer to be sure the voltage is correct. If there is moisture in the fuel lines, this can cause ignition delay. You should check and clean all filters, and you may have to drain the fuel storage tank, or at least the bottom portion where moisture and sediment settle. Electrodes may be dirty or improperly set, or the nozzle may need replacing.

If the flame does not burn cleanly, it may be due to a number of causes. The air adjustment may be closed. The pump pressure may be incorrect. The flue may be filled with soot. Check the igniter assembly. The burner nozzle may be damaged or too large, requiring replacement. Clean the nozzle, screen and igniters.

If the igniters do not arc and the pump is running, suspect an improperly adjusted or dirty igniter assembly. Remove the igniters, clean them, and set the gaps according to manufacturer's recommendations. Inspect them for cracks, burned electrodes or other damage requiring replacement.

If the igniters are okay and you still cannot get them to arc, suspect a malfunctioning transformer or improper line voltage. Line voltage can be checked with a voltmeter as outlined in Chapter 1. To check the transformer windings, attach the ohmmeter leads on the terminals of the secondary

Fig. 5-14. Check the pump pressure by removing plug (A) and installing a pressure gauge into the pipe fitting. Adjust pressure by removing cap nut (B) and turning the screw underneath. Follow the manufacturer's recommendations in setting the pump pressure; usually it will be about 100 psi. (C) Fuel line from pump to nozzle. (D) Fuel supply line filter.

winding. You should get a resistance reading (not infinity) if there is continuity through the winding. To check the winding for a short, attach one ohmmeter lead to one transformer terminal and the other ohmmeter lead to the metal housing. The reading should be infinity if there is no short circuit. If any reading other than infinity is given, the winding is bad. You can check the primary winding the same way. **Never perform ohmmeter tests without double checking to be sure the power is disconnected.**

Limit Control

The function of the limit control is to turn the burner off when very high temperatures are reached in the heat exchanger. If it is locked in an open position, it will not allow current to reach the stack control starting relays. If you suspect a malfunctioning limit control is shutting off current to start your oil burner, short across the limit control terminals. If the unit now runs, the limit control is bad and must be replaced.

Never reset the limit switch for a higher temperature setting. The setting should be around 250°F. A higher limit setting could cause a fire if something else were the cause of the malfunction. If the unit starts, replace the limit control.

Stack Control Safety Switch

Oil burners have a heat sensitive switch that turns off power to the pump if heat is not generated quickly enough after the pump starts. If the pump starts and then is turned off, this safety switch may have cut the pump off. The safety switch is usually part of the stack control, located near the flue.

First, suspect the igniters. If they are not operating, the heat sensing switch is doing its job. If the igniters light, the stack control may be malfunctioning. Normally this is caused by a clogged-up flue that insulates the stack control heat-sensitive probe to the point it cannot sense the heat from the burners.

Oil Delivery Problems

If oil is not reaching the burner, the burner can't work. Check all the unit's fuel filters to be sure they are clean. Suspect moisture frozen in the fuel delivery line. You may

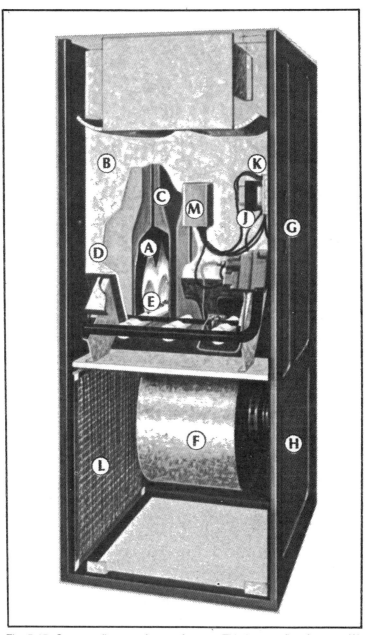

Fig. 5-15. Cutaway diagram of a gas furnace. This is an upflow furnace. (A) Combustion chamber. (B) Furnace vestibule. (C) Heat exchangers. (E) Burners. (F) Fan. (G) Gas supply and electrical wiring openings in furnace cabinet. (J) Transformer that reduces line voltage to 24 volts for control circuit. (K) Relay. (L) Filter. (M) Fan and limit control box.

have to drain the oil tank to remove moisture. Check for a clogged nozzle, or nozzle filter.

GAS BURNERS

Gas burning furnaces are among the most popular heating system installations today because of the traditional low cost of gas—particularly natural gas. However, these conditions are changing. Changes in government regulation of gas and chronic gas shortages means gas prices are escalating each year. As gas prices rise, it becomes more and more important to you to increase the efficiency of your gas furnace or boiler.

This section describes how you can tuneup and service the combustion system on a gas burning furnace, whether it burns natural gas or LP gas. These tuneup and servicing techniques are also applicable to gas-fired hydronic heating systems.

GAS BURNER OPERATION

Gas burners use one of two fuels. Because of its low price, *natural* gas is the most popular fuel used in gas burners. However, in some areas natural gas is not available, so LP gas is the type of fuel used in a gas furnace. Liquid petroleum gas is actually a term given to propane or butane, or a mixture of the two. From the serviceman's standpoint there is really very little difference in a gas burner using natural gas and one using LP gas.

The main difference in these types of fuel is their method of delivery. Natural gas is delivered by a local company through underground pipes. To get gas delivered to your home, you call the local gas company, and a connection between your house and the gas line hooks you into the system. The gas is delivered through the lines under enough pressure to force it into your furnace's burners. Liquid petroleum gas, on the other hand, is stored under pressure in a storage tank near your home. Part of the liquid vaporizes inside the tank, and this gas vapor is sent into your gas burners under pressure from the storage tank.

These are the primary parts of the gas burner's combustion system: gas valve, manifold, orifice, burners and the ignition system (either pilot light or electronic ignition). Figures 5-15 and 5-16 show the basic parts of gas furnaces and gas burners.

Chapter 2 introduces you to basic gas furnace operation and parts, such as the fan and heat exchanger. The central control unit of the gas furnace is the gas valve. The gas supply line is connected to this valve and gas is held here under pressure until released to flow into the burners.

The gas valve is a solenoid valve activated by the thermostat. When the thermostat calls for heat, the gas valve opens allowing gas to pass into the manifold that connects the

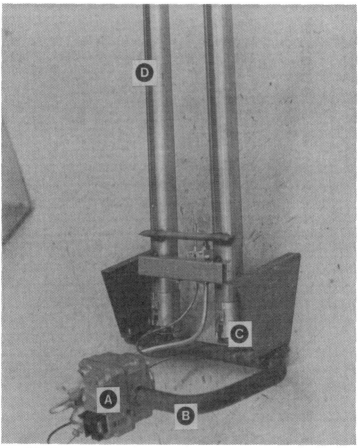

Fig. 5-16. This is the gas valve and burner assembly after it is removed from the furnace. (A) is the gas valve that is activated by the thermostat to send gas from the fuel supply lines into the burners. Once the gas valve opens, gas will travel through manifold (B). At the base of the manifold where the burners are attached at (C) is an orifice (one per burner) through which the gas must pass before going to the burner. The gas burns on the top surface of burners (D).

gas valve with the burners. From the manifold, the gas passes through an orifice where it mixes with air. This fuel air mixture now goes into the burners, where combustion takes place and heat is generated.

Depending on your furnace design, there may be two, three, four or more gas burners. Some furnaces have what is called a *spread-type* burner, in which the burner is circular-shaped rather than elongated.

PILOT LIGHTS, THERMOCOUPLES AND ELECTRIC IGNITION

Once the gas-air mixture begins flowing into the burners, there must be some device to light the fuel. Naturally, it is not possible for the homeowner to do this manually, so gas burners employ pilot lights or electric ignition systems to perform this chore.

The pilot light (Fig. 5-17) has been the lighting system used for years on virtually all gas furnaces and appliances. But rising fuel costs have made furnace designers and homeowners aware of the cost in fuel dollars of keeping a pilot light burning constantly. Thus, electric ignition systems are used on many new furnaces. But pilot lights are still the principal starting mechanisms on gas furnaces, because most gas furnaces were installed before electric ignition became so popular.

Pilot lights are located on the burner assembly, and they burn continuously. When gas begins flowing into the burners, the pilot light ignites it. Of course, if the pilot light is not lit, a dangerous situation would be created if the gas valve were allowed to open. The theromocouple is a safety device that keeps this from happening. The thermocouple extends into the pilot flame, and as long as the pilot light is lit, the thermocouple will allow the gas valve to open. If the pilot flame goes out, the thermocouple closes the gas valve so the thermostat cannot open it.

The pilot flame should burn with a bluish tint. If it burns yellow, the air opening for the pilot light is probably clogged and should be cleaned. The pilot light should be cleaned once a year as part of your regular pre-heating season tuneup. Cleaning the pilot is described a bit later in this chapter.

The pilot flame should not burn too high which will destroy the thermcouple in time and burn too much gas. There is an adjusting screw on the pilot light for this purpose. The flames from the gas burners should not touch the thermocouple, because they will destroy it.

Usually there will be a plate attached to your furnace giving exact instructions on how you should light your pilot light. The pilot light, of course, should be turned off during the summer cooling season, because it is a waste of fuel dollars to have it burning when there is no chance you will turn the furnace on. Therefore, the first time you turn on the furnace

Fig. 5-17. The pilot light and thermocouple are attached to the gas burner assembly. You can remove the pilot light and thermocouple from this assembly for cleaning. If the pilot light goes out, the thermocouple becomes cold and will not allow the gas valve to open.

each season, you will have to light the pilot. Also, you will have to light it any time it goes out. The furnace will not turn on unless the pilot is lit.

To light your pilot light, turn the thermostat down (so the burners won't light) and turn the gas valve to the pilot position. Your gas valve may have a knob you have to push down to get gas to flow to the pilot light. While pushing down on this knob, light the pilot, and hold the knob down long enough for the thermocouple to warm up and hold open the gas line to the pilot light. Then release the knob. If the pilot stays lit, turn the gas valve to the **on** position.

The pilot light may be difficult to reach with an ordinary match or lighter. One solution is to purchase a box of long fireplace matches that have wooden stems several inches long. You also can purchase a spring clip on a long heavy wire that will hold the match.

The pilot light may light but go out when you release the gas knob. Usually this means you have not held the knob down long enough to warm the thermocouple. If you have continued problems keeping the pilot light lit, there are several possible problems—dirty pilot light, defective thermocouple and gas line leaks. Tighten all gas line connections, and check for leaks with soapy water—particularly if yours is a new installation. Also on new installations, there may be air in the gas lines, which means you will have to continue attempting to light the pilot until the air has been removed. The gas burner tuneup portion of this chapter discusses how to clean a pilot light.

GAS BURNER TUNEUP

Once a year, you should give your gas burner combustion system a good tuneup and adjustment in preparation for the heating season. Cleaning the combustion system, inspecting it, and adjusting the air flow and gas pressure will make your gas burner operate at top efficiency throughout the heating season, as well as give it the service it needs for a long life.

Checking The Gas Valve

In proper operation, the gas valve will not allow any gas to pass through to the burners unless the thermostat is calling for heat. Sometimes, however, the gas valve may leak, which creates a hazard in the house. Once a year you should check your gas valve to be sure it is operating properly.

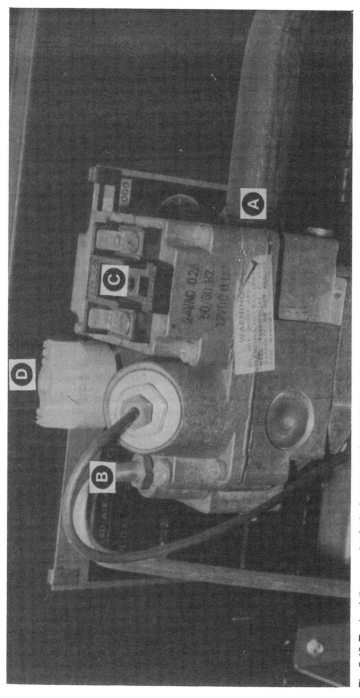

Fig. 5-18. To check the gas valve for leaks, remove the manifold and install a pressure gauge (A). (B) is the tube connecting the gas valve to the pilot light (C) is the solenoid control and the thermostat terminals. (D) is the control knob. Turn this knob to PILOT to light the pilot light.

To check your gas valve for leaks, remove the manifold from the gas valve and install a pressure gauge that measures pressure in water column inches. Usually, you will have to install an adapter fitting into the manifold connection before you can install the pressure gauge. Turn the gas valve to the **on** position. If pressure is shown on the pressure gauge, the gas valve is leaking. Be sure you have the thermostat wires disconnected from the gas valve, or the thermostat setting is low enough so the thermostat will not call for heat and open the gas valve. A leaking gas valve must be replaced, or it will allow gas to enter the living area. See Fig. 5-18.

Caution: Before disconnecting the fuel supply line from the gas valve for any reason, close the supply shut-off valve. This valve can be found at your furnace were the gas line connects to the furnace, or at the supply meter. With an LP gas supply, it normally is found at the supply tank.

Cleaning The Burners

Gas burners are just like any combustion system. When fuel is burned, a sooty carbon residue is left behind. Of course, the better your gas burners are in adjustment, the less of this residue there is. But over an entire heating season every gas burner will become carbon coated and sooty, which reduces the burner's efficiency. If left unattended, this carbon and soot will build up on the burners to the point that the burners will not heat at all.

You must remove the burners from the furnace heat chamber to clean them. The entire burner assembly will usually come out as one unit See Fig. 5-19. You will have to disconnect the pilot light tube at the gas valve, and you will have to disconnect the thermocouple connection. On some furnaces the burners alone will not come out, and you must remove the manifold with them. You may be able to disconnect the manifold from the gas valve to remove the burner assembly. Sometimes you will have no choice but to remove the entire burner/manifold/gas valve assembly as one. This requires you to disconnect all wiring and tubing connections from the gas valve. **Before disconnecting the fuel supply line from the gas valve, be sure to close the shut off valve.** Be sure to tag any wires or tubing removed so you can reinstall them in the proper places.

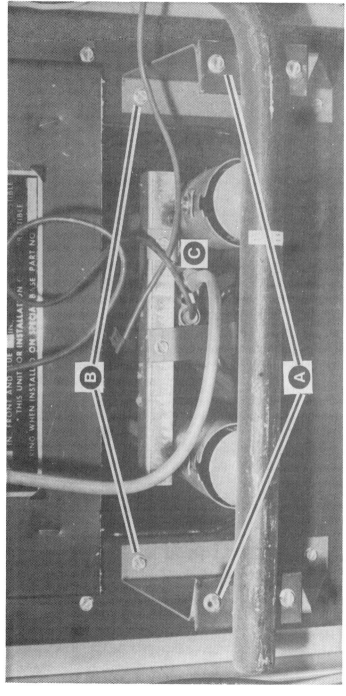

Fig. 5-19. To remove the burners and pilot light for cleaning, first remove screws at (A) to disconnect the manifold from the burner assembly. Remove screws at (B), and slide burners down and out. (C) is the pilot light and thermocouple.

The burners can be cleaned with compressed air or a cloth. Be careful that you do not bend or break the burners. You may also use a cleaning solvent as a last resort, but be aware that most of these are flammable. Allow the cleaning solvent to dry before lighting the burners. Never use water to clean the burners.

While you have the burners removed, you also should use a long-handled brush to remove soot and dirt that has built up on the combustion chamber walls.

Remove the orifices from the manifold and clean them with a solvent or an orifice cleaning bit (Fig. 5-20).

Cleaning the Pilot Light and Thermocouple

The pilot light should be removed from the burner assembly for cleaning. Soot buildup on either the pilot light or thermocouple can lead to malfunctions, but you must be careful in removing unwanted deposits. Carefully wipe off the thermocouple with a cloth, but do not use any cleaning solvents to clean it. If the thermocouple appears damaged, replace it while you have it out. Damage can occur if the pilot light is turned up too much or if the thermocouple touches the flame from the burners.

Clean the pilot light with compressed air or solvent. Take the pilot light apart, and inspect the small orifice inside it to be sure it is not clogged. This orifice may be cleaned with a pin or small wire.

When you reinstall the pilot light and thermocouple onto the burner, be sure you place the thermocouple directly into the pilot flame path. The thermocouple should not be located where it can be burned by the burner flames. After you reinstall the pilot light, install the burner assembly and reconnect all tubing and wires.

Adjusting the Gas Pressure

If you have access to a pressure gauge that measures pressure in inches of water column, the next step in your annual gas burner tuneup is adjusting the gas pressure. You might consider the purchase of such a pressure gauge at your heating system supply house, especially if you will be tuning up several gas furnaces each year. If you do not have such a pressure gauge and do not want to purchase one, you should

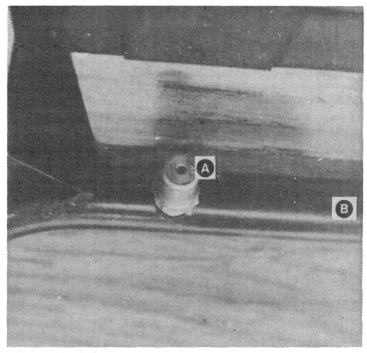

Fig. 5-20. Once you have removed the burner assembly for cleaning, you will find an orifice (A) attached to the manifold (B). An orifice extends into each burner, metering the flow of gas from the manifold into the burner. Unscrew the orifice from the manifold and clean it with a cleaning solvent, compressed air or an orifice cleaning bit.

not attempt to set the gas pressure. If you do not have a pressure gauge and you suspect gas pressure problems, contact your local gas supplier or heating serviceman. The gas supply company will normally check and adjust the gas pressure for you, often without charge. Before you can set the gas pressure yourself, you need the furnace manufacturer's recommendations of gas pressure, which are usually in the furnace instruction manual.

Gas pressure problems should be suspected if you have made all tuneup adjustments listed in this section (including air adjustment discussed shortly), and the flame does not burn properly. If you have cleaned the burner, adjusted the air and the flame is too large or too small, suspect incorrect gas pressure.

Fig. 5-21. Most furnaces have two pressure regulators to set the gas pressure. This is a supply tank pressure regulator for an LP gas furnace. The pressure regulator is beneath the cap. Similar pressure regulators are found at the home meter of natural gas systems. Before adjusting the supply tank pressure, contact the fuel supply company. They may send a company technician to adjust this regulator at no charge. In any case, you usually need the company's approval before you can adjust the supply pressure regulator.

If it is available, you should consult the furnace instruction manual before you begin checking the gas pressure. Usually this will tell you where the pressure regulators on your furnace are located, and it will tell you what the correct pressure is.

To check the gas pressure, first turn the gas valve to the off position. Install the pressure gauge into a plug found either on the gas valve near the manifold or at the end of the manifold. Sometimes you will have to remove the orifice of one burner from the manifold and insert the gauge into the manifold in place of the orifice. After installing the gauge, turn the gas valve to the on position, relight the pilot light and set the thermostat so the burners will light.

If the gauge indicates the gas pressure is incorrect, you will have to adjust the pressure at one of the pressure regulators on the fuel line system. One pressure regulator is normally located at the supply source—the storage tank for a LP system or the gas meter for a natural gas system. See Fig. 5-21. As a rule, gas companies have policies against unauthorized persons setting the gas pressure at the supply source. If you contact the gas supplier, he will normally make necessary gas pressure adjustments for you.

Some furnaces have only one pressure regulator—the supply line regulator. But most have one other pressure regulator located either in the gas valve or as a separate unit in the fuel line just before the gas valve. See Fig. 5-22.

The best way to proceed in adjusting the gas pressure is to begin with the pressure regulator at the gas valve. If adjusting this regulator does not achieve the gas pressure you need, go next to the fuel supply pressure regulator. Contact the gas company before adjusting this regulator. If they approve it, go ahead and adjust the supply pressure regulator. Then return to the gas valve regulator to make exact adjustments.

Adjusting the Air Mixture

If you adjust the gas pressure, you should adjust the air mixture afterward. Before you adjust the gas pressure, how-

Fig. 5-22. Some furnaces have only the supply pressure regulator. Most, however have at least one more pressure regulator either at the gas valve or in the supply line just before the gas valve. To adjust the gas pressure, remove the cap and adjust the screw underneath.

ever, see if flame problems clear up by adjusting the air mixture. Try this especially if you do not have access to a pressure gauge.

The correct gas flame is blue. A yellow flame indicates a lack of air. If the flame burns off the burner top, it is an indication of too much air flow and gas pressure. If reducing the air flow does not correct the problem, the gas pressure may be set too high.

Figure 5-23 shows the air adjustment for one type of gas burner. Most burners have similar adjustment mechanisms that are easy to adjust. Usually all you have to do is loosen a set screw and rotate or slide a device to open the air inlet.

Adjusting the Pilot Light

The pilot light should be blue, just like the flame at the burners. A yellow pilot light indicates a dirty pilot light assembly or a clogged line. The pilot light should be just large enough to reach the thermocouple and to light the burners. If the pilot light is too large, it will damage the thermocouple. The size of the pilot light can be adjusted by rotating a screw at the gas valve where the pilot light gas supply tube connects to the gas valve.

Visual Inspection

An important part of any tuneup is a thorough visual inspection to locate any potential future trouble spots. Clean out the furnace control area and the furnace fan, as described in Chapter 4. Be sure all tubing connections are tight and free of gas leaks. Inspect all wiring and terminals to be sure screws are tight and wires are secure.

GAS BURNER TUNEUP CHECKLIST

1. Check gas valve for leaks using pressure gauge.
2. Clean burners and orifices.
3. Clean pilot light; inspect thermocouple.
4. Check and adjust gas pressure using pressure gauge.
5. Adjust air mixture until flame is blue.
6. Adjust pilot light.
7. Visually inspect all connections, tubing and wires.

Fig. 5-23. Opening and closing the air inlet (A) adjusts the amount of air flow to the burners. A set screw (B) holds the air adjustment mechanism in place. The flame should be a blue color. Yellow indicates too little air, while a flame burning off the burner indicates too much air.

TROUBLESHOOTING AND SERVICING GAS BURNERS

Sometimes you will have a problem with the burners not lighting. Although there may be a number of causes, including a defective or dirty pilot light and dirty burners, another cause is a gas valve locked closed. A coil in the gas valve opens the valve when the thermostat calls for heat. If this coil is defective, the gas valve will not open.

You should check this coil by performing a continuity check (see Chapter 1). Remove one of the wires from the thermostat terminals on the gas valve. Place the two ohmmeter leads on the terminals. The meter should read continuity.

Check the voltage to the coil. Depending on what your furnace design is, this voltage will usually be 24V. Finally, remove the manifold and install a pressure gauge on the manifold connection, as you would if you were checking for a leaking gas valve. Connect the thermostat and turn it on. Be sure the pilot light is on. Now, the gas valve should open and the furnace would light if the gas valve were connected and working properly. If no pressure reading is given, double-check to be sure the pilot light is on and the thermostat is connected and calling for heat. Be sure the gas valve is turned to the position.

No pressure reading means the gas valve is defective and will have to be replaced. If pressure is shown, there may be blockage in the manifold or in the orifices that is not allowing the gas to get into the burners.

Transformers

If transformer problems exist, they can halt the entire furnace. The gas furnace has one transformer that reduces the line voltage from 120 to 24 volts for the controls and thermostat circuit. If you suspect transformer problems, you should check the voltage or the primary winding and secondary winding with a voltmenter. Check the windings for continuity and for a short following the procedure outlined in the section on oil burners and transformers.

If the burner will not light, the pilot light may be out, the power supply to the furnace may be disconnected, there may be a loose thermostat wire, the gas valve may be turned off, or the gas valve could be locked closed.

Other possibilities are insufficient voltage on secondary side of transformer, a loose transformer connection, a defective themocouple and a limit switch locked open.

With a yellow burner flame, check for low air adjustment, dirty burners, dirty orifices and low gas pressure. If the burner explodes on lighting, the pilot light may be set too far away from the burner, dirty or set too low. There may be a leaky gas valve or improper gas pressure.

Causes for the pilot light continually going out include air in fuel lines, LP gas tank out of fuel, defective thermocouple, pilot light adjusted too low, thermocouple not in pilot light flame path and a dirty pilot light.

ELECTRIC FURNACES

Electric furnaces generally require the least maintenance, servicing and tuneups of any of the three most popular heating fuels. This is primarily because electric furnaces do not have a combustion system that must be kept in tune to keep the furnace operating at peak efficeincy.

This does not mean that electric furnaces require *no* maintenance, but many of the routine maintenance and tune-up procedures are discussed elsewhere in the book and are applicable to almost all furnaces. Part of your annual furnace tuneup should include cleaning the fan, cleaning the humidifier, checking your thermostat's operation, cleaning or replacing filters and a thorough visual inspection of the wiring and wiring terminals. Performing these tuneups tasks is discussed in Chapter 4. A regular maintenance program that includes changing or cleaning furnace filters periodically and cleaning the furnace fan once a year will keep your electric furnace operating at near-peak efficiency.

SERVICING THE ELECTRIC FURNACE

Although there are not a lot of "once-a-year" tuneup chores that apply specifically to an electric furnace, there are a number of procedures for servicing that are fairly routine. If you own an electric furnace or are occasionally called upon to service one, you should be familiar with these common electric furnace problems and repair techniques.

As explained in Chapter 2, the electric furnace produces heat by the use of electrical current through metal elements that have a definite resistance to electrical flow. A fan blows air over these elements to pick up heat and take it into the living area. Because there is no combustion process going on in an electric furnace, there is no combustion chamber or heat exchanger. The air moving through the furnace blows right through the elements and comes in direct contact with them.

You will find electric furnaces in any kind of installation: upflow, downflow or horizontal. The parts of any electrical furnace may be arranged to form any of these installations.

Electric Furnace Operation

Because the electric furnace does not have a combustion process, it operates quite a bit differently than gas or oil burning units. The primary parts of an electric furnace's heating cycle are the thermostat, sequencers and heating elements. The heating elements, as already described, are made of electricity-resisting metal that will heat up when electricity is sent through it. But all the elements do not come on at once. A typical electric furnace will have four heating elements, and they come on one at a time.

Turning the elements on one at a time is accomplished by *sequencers*. These sequencers are activated by the thermostat so that when the thermostat calls for heat, the no. 1 sequencer (controlling the no. 1 element) is activated. When the no. 1 sequencer is activated, it sends current to the no. 2 sequencer, after a slight delay. This activates the no. 2 sequencer and the no. 2 element and sets up the no. 3 sequencer, and so on. If a two-stage thermostat or an outdoor thermostat is wired into the electric furnace, the line between sequencer no. 2 and sequencer no. 3 will probably be taken to that thermostat. The purpose of this is to keep elements no. 3 and 4 from coming on unless they are really needed. Chapter 11 includes a discussion of outdoor thermostats and two-stage thermostats.

The electric furnace usually has a heat-sensitive fan switch that turns on the furnace fan after element no. 1 heats up. The furnace also has safety limit switches that switch the furnace off if the elements get too hot, as would be the case if the fan did not come on or there was a restriction in the duct system. The safety limit switch keeps the elements from burning out.

Wiring and Overload Protectors

Almost every electrical device will occasionally experience overloads in the electric circuits, which causes the overload protectors (either circuit breakers or fuses) to open the circuit for safety. The electric furnace is no exception. There are two types of electric furnace wiring, and the overload protection system for your furnace will depend on how the furnace is wired.

The electric furnace is normally on a 240 volt circuit. Remember to use caution in working with 240 volt circuits, because a shock from such a circuit can kill you. One way to wire the furnace is to have each element wired separately into the household breaker box with its own breaker. In this wiring system, at least one of the overload protectors for each element circuit is located at the breaker box.

The second type of wiring system has the entire furnace on one large circuit leaving the breaker box. This circuit then is split at the furnace into separate circuits for each element. In this type of wiring, it would not be unusual to have the furnace wired on a 240 volt circuit with a 125 ampere breaker. If the same furnace were wired with individual breakers for each element, there would be perhaps four 24 volt circuits with a 30 ampere breaker on each circuit.

With the second type of wiring system (one 125 ampere circuit), we must have overload protectors at the furnace for each element's circuit. This is because the single 125 ampere breaker is so large that if one of the element circuits were to draw too much current due to a malfunction, the breaker would be too large to open the circuit before damage were done. So each element circuit has its own overload protec-

Fig. 5-24. The pointer shows the fuses in the furnace that protect each element circuit. If the element circuit overloads and breaks one of these fuses, you will have to replace it. It is a good idea to keep at least two of these fuses on hand for spares.

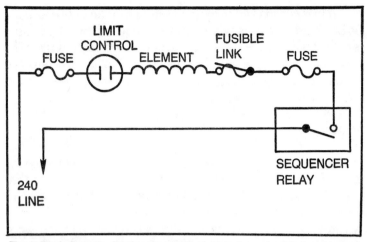

Fig. 5-25. A diagram showing the overload protectors normally found in each electric furnace element circuit. If the element is wired on its own circuit from the breaker box, the fuse nearest the power line will be found in the breaker box and not on the furnace. The sequencer relay is activated by a 24 volt circuit that is not shown.

tor, such as a fuse, located on the furnace for easy access if one needs replacing (Fig. 5-24). If the elements were 20A elements, these fuses might be 30A fuses—one for each element circuit.

On some furnaces, you will find a second fuse of the same type for each element, and/or a bimetal disc or link also in each element circuit as shown in Figs. 5-25 and Fig. 5-26. The disc overload protectors are to protect the circuit when the element overheats.

Checking the Element Circuit

If you suspect a furnace element is not working, use a clip-on ammeter around a wire on one terminal for each element to see if current is flowing through the element circuit. If your furnace has a two-stage thermostat or an outdoor thermostat that would keep some of the elements off, be sure the thermostat contacts are closed or the contacts have been shorted across so that all elements will turn on. If you don't have an ammeter, you can make continuity checks with an ohmmeter on each circuit (with the power turned off). This method would be a little slower way of isolating the defective element circuit.

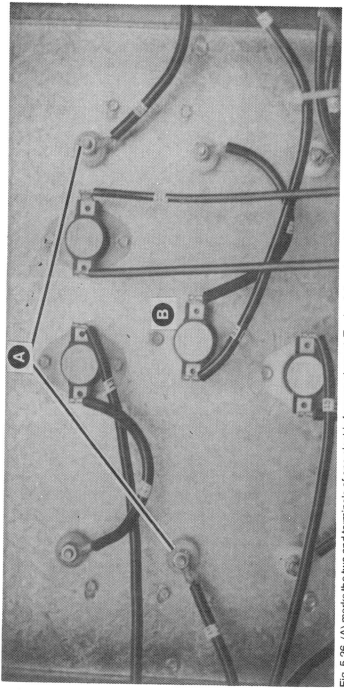

Fig. 5-26. (A) marks the two end terminals of one electric furnace element. To check the element to see if it is good, place the ohmmeter leads on these terminals. (B) is the fusable link disc overload protector that opens the circuit if the element gets too hot. When checking an element, be sure to remove one lead from the element terminal to stop any feedback through another circuit.

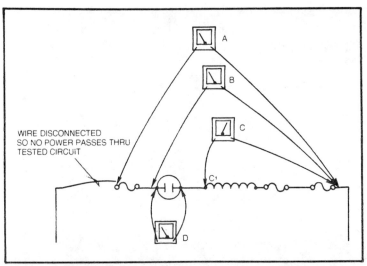

Fig. 5-27. When you suspect a defective element circuit, first isolate the defective circuit, and then begin running continuity checks on the circuit with an ohmmeter until you isolate the trouble spot. Begin by attaching the ohmmeter leads at the beginning and end of the circuit as at A. An open circuit is shown on the meter by a no continuity reading (infinity). Now, proceed along the terminals of the circuit. At B, again no continuity is shown. But at C, there is continuity, meaning the remainder of the circuit is good. Now, proceed backwards with the ohmmeter, testing each part up to C1, to see if it is good. D shows no continuity through the limit switch, meaning it is the defective part.

Once you have located the element circuit that is inoperative, you should make continuity checks across the circuit until you isolate the defective part. Normally, this will be a fuse or circuit breaker that is open, but in some cases the element will be the problem. Figure 5-27 shows how to proceed with a continuity check of the element circuit.

To check an element for a suspected short, place one ohmmeter lead on an element terminal and the other lead on the metal furnace case. If any continuity reading shows up, the element is shorted to the case.

Transformers

The control circuit of the electric furnace is 24 volts, and this circuit is connected to the thermostat and sequencers. A transformer steps down the voltage coming in from the line from 240 volts or 208 volts to 24 volts. For the furnace to operate properly, the transformer leads connected to the power line must match the incoming line voltage. The transformer primary circuit has leads for 208 and 240 volts line. If your power line is 208 volts and the line is connected to the 240 volt terminal of the transformer, chances are the sequencers on the furnace will not operate.

The transformer's primary circuit is protected by its own 15 amp fuse. If you suspect transformer problems, be sure to check this fuse.

Transformer Checks

To check the voltage reaching the furnace and the transformer, place the voltmeter leads on the terminals R and C on the furnace terminal board. Power should be turned on. If you do not get a voltage reading, you are not getting power to the furnace.

To check the continuity of the transformer windings, and to check for short circuits, follow the procedure outlined in the section on oil burners and transformers.

Furnace Fuel Conversions to Cut Heating Costs

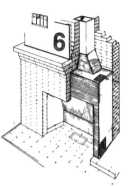

Most of this book is devoted on how you can save energy and money by leaving your furnace pretty much as it is. This chapter, however, discusses how to go about achieving lower fuel costs by changing your basic furnace equipment to accept a different fuel. In some cases, this requires complete replacement of the furnace. In others, you have to change only the combustion system, but the furnace body remains in place. With others, such as when adding a supplementary wood furnace, you have to install the additional unit and make some duct system changes, but the basic combustion system you now use will remain intact.

Changing the fuel of your furnace may seem like an overly ambitious task, but in many cases it is really quite easy. And the results can be staggering! Adding a small wood furnace to your heating system as a supplementary heater or installing a heat pump coil in an electric furnace—two of the most common modifications—can save you one-third, one-half or more on your heating bill.

There is no question, however, that changing the fuel your furnace uses is a big step and one that you should not take lightly. It also will require a considerable investment, but if you have calculated your fuel costs correctly, the investment will pay off in the long run.

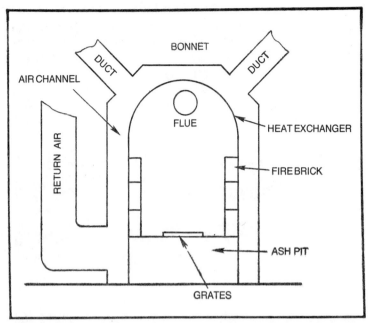

Fig. 6-1. This is a cross-section diagram of an old style coal furnace. The air entered the furnace at the return air duct and was free to move through the air channel next to the firebox. The heat from the firebox warmed the air inside the air channel surrounding the firebox. The heated air rose to the bonnet and out through the two or three large ducts that connected the furnace with the living area above. All air movement was by natural convection air currents.

Your first step, of course, is to do fuel cost analysis of the various fuels you are considering switching to, as described in Chapter 3. You have to remember that any future savings by making a fuel conversion will be offset at least partially by the cost of changing over. Of course, if you can get a good deal on selling your current furnace—if you do have to replace it to change your fuel—that makes the offset considerably less. Also, if your current furnace is nearly worn out and ready for replacement anyway, the offset against future savings is much less. The remainder of the chapter discusses how you can perform some of the more common furnace conversions.

COAL FURNACES

Chapter 2 introduces you to what we will call the *old style* coal furnace. These furnaces were quite popular earlier in this century and were the first step between the old wood stoves and today's modern heating systems. Old style coal furnaces did not have a fan; nor did they have the modern duct systems so common today. Instead, they depended on the natural tendency of warm air to rise to produce convection air currents that circulated the heat through the house. The furnace was always located in the basement under the house, and the furnace was connected to the living area with two or three very large ducts. Diagrams of the old style coal furnace installation appear in Chapter 2. Figure 6-1 shows the parts of this furnace. Most of these furnaces have long since been replaced or converted over to a forced-air system, but occasionally you will find a coal furnace that has not been modified.

A newer style forced air coal furnace is shown in Fig. 6-2. This diagram also shows the wiring and controls necessary to modernize an old style coal furnace. The forced air furnace has a fan and a duct system for more efficient air distribution through the house. It also has a stoker, which automatically augers coal into the firebox as it is needed. (The old style coal furnaces were hand fed).

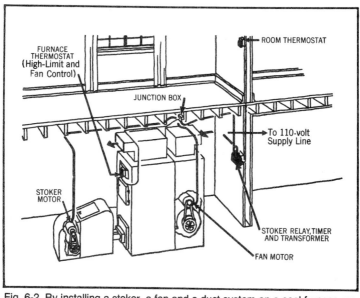

Fig. 6-2. By installing a stoker, a fan and a duct system on a coal furnace, we have a forced air coal furnace. Many of these more modernly designed coal furnaces have since been converted to gas or oil furnaces because of the convenience in handling that type of fuel. In this diagram you can see the controls necessary if you convert an old-style coal furnace to a stoker-fed forced air furnace. The controls needed are the limit switch, fan control, and the stoker relay (courtesy U.S. Dept. of Agriculture).

There aren't very many residential coal furnaces left around the United States anymore, although at one time coal was a very important home heating fuel. The primary reason that few coal installations remain is the difficulty in handling coal. The stoker usually must be loaded manually, and a large bulky coal storage area must be set aside to hold a supply. Gas, oil, and electricity are simply more convenient although coal traditionally has been the cheaper fuel.

Most coal furnace conversions will involve installing a stoker and fan on an old-style coal furnace or converting a forced-air coal furnace to a gas or oil burner.

To add a stoker and fan to an old style coal furnace, you will have to add a stoker in place of the ash grates. A low-voltage electrical line can be used to wire the stoker motor to the thermostatically controlled stoker relay. The fan is added at the base of the furnace so it will force air upward through the air channel and into the duct system. The return air duct is connected to the fan housing. This fan must be controlled by a limit switch and fan control as described in Chapter 4 in the section on fan controls.

To convert a coal furnace to a gas or oil furnace, there are a number of conversion kits available. These were quite popular in the past when there were a large number of coal furnaces in homes. One such kit, a gas burner, is shown in Fig. 6-3. A furnace should have a fan and a duct system before the gas or oil burner conversion kit is installed.

Although there has been renewed interest in coal heat recently with the increasing prices of fossil fuels there has so far been little interest shown in wholesale conversion of furnace systems entirely back to coal. The difficulty in handling coal is partly responsible for this, as is the general lack of residential coal suppliers. There has been some interest in using coal as a supplementary heat. This keeps the residential homeowner from having to depend totally on a coal supply for his heat source. Two systems are currently available for using coal as a supplementary heating fuel. One is a combination furnace that burns coal or wood and a secondary fuel that usually is oil or gas. Usually the fossil fuel is used only as a back-up fuel in the furnace. The combination furnace is capable of producing the entire heat load of the house by either coal or the "secondary" fuel. Combination furnaces are discussed in

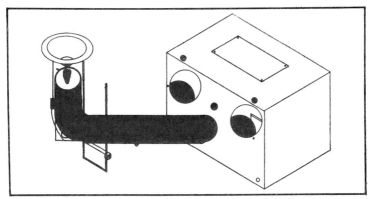

Fig. 6-3. This is a conversion gas burner that installs in the firebox of a coal furnace to convert it to a gas burner. The primary reason for converting coal furnaces to gas or oil burners has been convenience in handling those fuels, not necessarily lower fuel prices. Similar oil burner conversion kits are also available.

more detail in Chapter 9. Their installation requires the removal of your current furnace so the combination furnace can be put in its place.

The second method of taking advantage of coal as a heating fuel is to install a coal heater in line with the duct system, in the same manner as you would install a wood heater. In fact, many of the wood units that can be added into the furnace duct system for supplementary heat come in coal-burning models. In such an installation the coal heater is a supplementary heater only. It is not intended to be the primary fuel for the home's heat load, and it requires manual feeding. The installation of a four-to-five room size coal heater/furnace is discussed in the following section. You also should see Chapters 8 and 9 for a discussion of how to install these units.

ADDING WOOD AND COAL HEATERS TO YOUR HEATING SYSTEM

A lot of interest has been shown in recent years in utilizing the potentially cheaper fuels of wood and coal in the heating system. One way to do this, of course, is to install a fireplace or a several-room wood or coal stove in the home's living area. The disadvantage to this is that poor heat distribution is usually a problem. The one or two rooms close to the stove will be quite hot, while little of the wood or coal heat reaches the far bedrooms of the home.

This problem can be solved by connecting what has been called an additional furnace or converter furnace into your present duct system. These units are actually similar to large wood stoves with a five room or so capacity. But they are enclosed in a metal jacket so the hot air they produce can be directed through a duct and connected to your present furnace duct system. By doing this, you utilize the furnace's fan and duct system to achieve an even distribution of heat throughout the house. The wood or coal heat produced by the additional furnace provides supplementary heat that reduces the amount of gas, oil or electricity you must use. It will not eliminate the use of these primary fuels, however, because the primary furnace must be heating before the furnace fan will turn on. They do, however, reduce the amount of the primary fuel you will have to use.

Some add-on wood and coal heating units are available with their own fan and thermostatic controls that can distribute wood-heated air through the duct system independently of the primary furnace. The fan, however, must be large enough to move air through the entire duct system. Thus, if your present furnace has a fan rated at 1000 cfm (cubic feet per minute), the fan on the wood or coal add-on furnace must be 1000 cfm to distribute wood-heated air to all parts of the house. You can check your furnace fan rating on the furnace nameplate or in the owner's manual.

A wood or coal add-on furnace capable of distributing heat independently of the primary furnace will be more efficient in its use of wood and will save you money if wood or coal are cheaper fuels than your primary fuel. This is because the wood or coal heat can be distributed without using the primary fuel to turn on the furnace fan. You can expect, however, to pay more money for an add-on unit with these capabilities. Heating systems with wood or coal heaters installed with their own fans and controls actually would use wood or coal as their primary fuel. Then, if the wood heater were unable to take care of the entire heat load of the house, the "main" furnace using gas, oil or electricity would turn on.

Installing these add-on furnaces is discussed in two more sections of this book. Chapter 8 shows how you can connect the duct system for these units. Chapter 9 discusses the types of heaters available and shows you some installation diagrams.

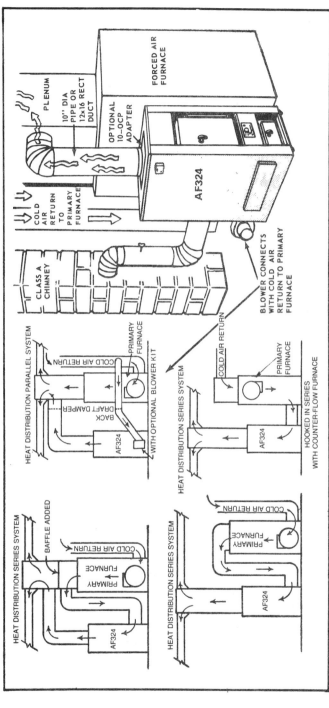

Fig. 6-4. These diagrams show you several of the many ways an add-on wood or coal heater can be installed in line with your present furnace system to distribute the wood and coal heat evenly throughout the house. Before purchasing one of these units, compare the features of the different models and examine your home's capabilities for installing them. Check with local authorities to see if there are restrictions as to installation of wood or coal-burning units (courtesy Monarch Ranges and Heaters, Division of the Malleable Iron Range Company).

There are a number of different ways these units can be installed with your present heating system, and some of them are shown in Figs. 6-4 and 6-5. You can connect the additional wood heater in series with your present furnace so that air leaving the furnace enters the wood heater before entering the duct system, or you can connect the furnace so that a portion of the cold return air is diverted into the heater and sent separately into the duct system.

Factors to Consider

There are several things you must consider before installing one of these add-on units. You must think about where you will place the additional wood heater and how you will connect it to your present furnace. There may be room in your present furnace area to place the wood unit alongside the present furnace. Or you may have to place the wood unit in another area, such as a garage, and duct the hot air from the wood unit to the furnace. You must be careful not to place it too close to any combustible materials, and follow the guidelines in Chapter 9 on safe placement.

You must also check into local building and fire codes before you install. Some are very strict in their restrictions on wood and coal-burning heaters.

Combination Furnaces

Another way you can obtain wood or coal heat without having to depend on it entirely is by installing a combination furnace. A combination furnace, as described in Chapter 9, is a furnace that burns two fuels. Usually these furnaces are capable of burning wood or coal in combination with either gas or oil. These furnaces allow you to use wood or coal as your primary fuel, while retaining gas or oil as a back-up fuel if the fire gets too low or the wood or coal heat will not take care of the home's heat load.

Because the combination furnace requires the removal of the house's present furnace, combination furnace installations will be most popular in new home installations, as replacements for worn-out furnaces, and as substitutes for present furnaces where there is not enough space to install an add-on furnace described above. Replacing your present furnace with a combination furnace is discussed later in this chapter in the section on replacing and installing furnaces.

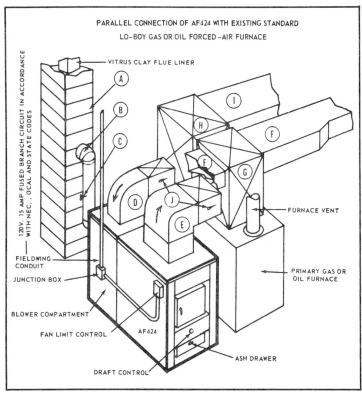

Fig. 6-5. This diagram shows how a large add-on furnace would be connected. This furnace model has its own blower and thermostatic controls, so it does not require the primary furnace to run to distribute heated air throughout the house (courtesy Monarch Ranges and Heaters, Division of the Malleable Iron Range Company).

GAS FURNACES

The most likely conversion with a gas furnace is to convert an LP gas furnace to burn natural gas. In most areas, natural gas is considerably less expensive than LP gas, but in recent years natural gas users have become quite vulnerable to lack of available supply. A homeowner might be interested in a natural gas to LP gas conversion if he lived in an area of chronic natural gas shortages and wanted to convert to a fuel with a more dependable supply, even if it is more expensive. Of course, the reverse is also likely. A homeowner unable to get natural gas when his furnace was installed might want to convert his LP gas furnace to natural gas after natural gas once again became available.

Essentially, LP to natural gas conversions, and vice versa, are simple to do since these furnaces are very similar. The orifices on the gas manifold will have to be changed on some furnace models and so will the gas valves. Your gas supplier and heating supply shop should be able to help you determine just what you will need to convert your particular furnace. Chapter 5 in the section on Gas Furnace Tuneup discussed orifices and gas values.

Certainly before you make any changes in the type of gas you use, you will have to contact both LP gas suppliers and natural gas suppliers to arrange to disconnect your present fuel supply and reconnect a new one.

In some regions where oil is less expensive to burn than LP gas, the LP gas users will want to change over to oil. This is a conversion that cannot usually be done without replacing the existing gas furnace with an entire oil furnace. The one exception is coal furnaces that previously have been converted to gas burners, as described earlier in this chapter. Usually it is a fairly simple matter to remove the gas burner conversion assembly and replace it with an oil burner conversion assembly.

Gas furnaces themselves cannot be changed to oil furnaces because of the design limitations of the typical gas furnace. The gas burner's heat chamber is not heavy enough for an oil burner, and there is usually is not enough space in the gas burner's combustion chamber to fit an oil burner. Therefore, the best solution is replacement, if you are set on converting your gas system to oil. Converting gas hydronic systems is exactly the same as gas furnaces.

OIL FURNACES

If you wish to change your oil burning furnace over to a gas furnace to take advantage of easier fuel handling or lower fuel costs you may be able to convert your furnace without replacing it. Unlike the gas furnace, which is generally not designed with the capability to install an oil burner, the oil furnace frequently can be converted to burn gas.

You will have to install a gun-type gas burner conversion assembly as shown in Fig. 6-6. This assembly includes a gas valve attached to the gas burner. You will have to connect the gas supply line to the gas valve, and the thermostat wires

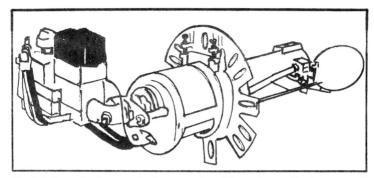

Fig. 6-6. A gas-fired conversion burner such as this can be used to change an oil furnace into a gas-burning furnace. The assembly includes the burner and gas valve. To install it, you remove the gun-type oil burner from the furnace and put the conversion burner in its place.

presently used are connected to the terminals on the top of the gas valve solenoid switch. Chapter 5 discusses the gas valves and solenoids in detail.

To install the conversion system, you simply remove the gun-type oil burner from your oil furnace, and install the conversion gas burner in its place. Then connect the gas supply line and the thermostat connections. Be sure to contact your gas supplier to arrange a gas supply before making this change. Converting an oil hydronic system is the same as an oil furnace.

WHEN INSTALLING A NEW FURNACE

In some cases it will be an economical choice to install a replacement furnace for your existing unit. Generally, this will be the case only when heating costs with an alternate fuel are extremely low (such as having your own wood supply) or when your old furnace needs replacing anyway. Also, if you can get a good deal on reselling your old furnace, a replacement may be an economical choice.

There are several things you must consider when you elect to replace your present furnace with one that burns a different fuel. First, you may have to make some modifications to your house. Many furnaces are installed in very small furnace rooms that have little room to spare. If you select a large, bulky furnace to replace it, you will either have to build a larger furnace room or install the furnace in another location and rearrange the duct system to it.

You will have to check your present furnace fan's capacity in cfm and be sure the replacement furnace fan has about the same cfm (cubic feet per minute). Thus, if you check your present furnace's nameplate or owner's manual and discover the furnace has a 1000 cfm capacity, your replacement furnace fan must also be 1000 cfm. If it is not, you may have to change the size of some of the ducts in the duct system to accommodate a different size fan.

If you install a replacement furnace with a fan that is rated at more cfm than your present fan, you will have a lot of air noise in the duct system. Sometimes this can be corrected by slowing down the fan speed. If your replacement furnace is rated at too low a cfm, the fan will be unable to push enough air through the duct system and many rooms will not receive any heat because of poor air distribution. Sometimes this can be corrected by increasing the fan speed.

Usually you will have to do some ductwork on the return air duct or the warm air plenum to make them connect to the new furnace properly. If the replacement furnace will be in about the same location but is taller or shorter than the old furnace, the plenum chamber or return air chamber above the furnace can be shortened or lengthened to reach the furnace.

To install the replacement furnace, you have to disconnect the old furnace from all fuel lines, wiring and the duct system. Usually the furnace will slide out after you push up the plenum chamber or return air chamber above the furnace.

The replacement furnace is slid into place and the ducts are reconnected. Unless you are installing an electric furnace as a replacement, your existing thermostat lines can be used for the new furnace. If you are installing air conditioning as part of the new furnace, however, and there was none before, you will have to run lines for a new heating-cooling thermostat to replace the old heating-only thermostat. You will have to run the fuel lines and connect the new fuel system.

If you already have air conditioning installed and you want to replace your existing furnace, you can sometimes avoid disconnecting the air conditioning coil and all the refrigerant lines. To do this, you must disconnect the coil case from the heating part of the furnace. If you have an upflow furnace and the coil case is located above the heating section, you should support the coil case by tying support wires to the ceiling.

Once the coil case is disconnected, remove the old heating section and slide the new one in place. This procedure will work, however, only if the replacement furnace is about the same size and configuration as the old furnace. If the new furnace is considerably different, you will hve to install your old coil into a coil case in the new furnace.

ADDING HEAT PUMPS

The heat pump has received a lot of attention in recent years because of its energy saving potential. Especially when it is compared to electric resistance furnaces, heat pumps usually can save quite a bit on your fuel bill. Chapter 10 discusses heat pumps and their operation. In this section we are concerned only with how you can add a heat pump on to your existing furnace without having to purchase an entirely new furnace system.

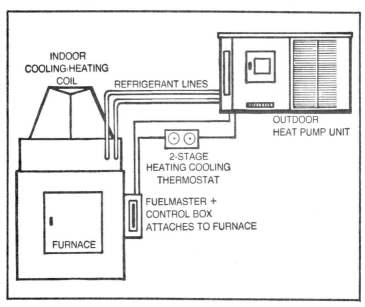

Fig. 6-7. This is an add-on heat pump system installed on an existing upflow furnace. The indoor heating/cooling coil is installed on top of the existing furnace assembly, and when fully installed, a coil case will surround this coil. The outdoor heat pump unit is connected to the new coil with refrigerant lines. When the controls and two-stage thermostat are installed, the heat pump is the primary heating system for the home and the existing furnace burners become the heat pump's supplementary heaters. This type of add-on heat pump system is especially good for those who have furnaces a few years old but who want to take advantage of energy savings with a heat pump.

The usual new heat pump installation includes an indoor coil, outdoor coil, a compressor and banks of electric resistance heating elements for supplementary heat. If you are a homeowner who now has a furnace—an electric furnace for instance—and do not have central air conditioning, a heat pump installation might be something you should consider if you are planning to add central air conditioning. But you already have your present furnace, and what do you do with it? Well, every heat pump needs some sort of supplementary heating system, and you can keep your present furnace and use it for the supplementary heat. Maybe you don't have an electric furnace, but instead have a gas or oil furnace. Those will usually work, too, depending on their design.

So if you already have a furnace installed and want to install a heat pump (you first must make a fuel cost analysis as in Chapter 3 before deciding whether you will save money with a heat pump installation), you can purchase the heat pump coils, controls and compressor alone. You don't have to purchase the electric resistance heaters, the indoor coil cabinet, or the duct system, because you already have those installed.

Several manufacturers make these "add-on" or piggyback heat pump systems for installation in existing furnaces. Thus, the homeowner's present furnace system supplies the supplementary heating unit, the furnace cabinet/fan and the duct system, while the piggyback heat pump kit supplies the indoor and outdoor coils, refrigerant lines, compressor and controls. Fig. 6-7 shows one of these systems and how it attaches to the existing furnace.

Installation of Indoor Coils

If you have an upflow furnace, the heat pump's indoor coil will be installed on top of the present furnace case, and a coil case will have to be installed around the coil. To make room for the coil and coil case you will have to cut away part of the plenum chamber above the furnace or, in a basement installation, reposition the plenum and main duct to make room for the coil case.

If you have a downflow furnace, the indoor coil will be installed underneath the present furnace next to the plenum chamber. A coil case will have to be installed here to house the coil. To make room for the coil case, you will have to cut away part of the return air duct at the top of the downflow furnace and raise the furnace up to install the coil case.

Wiring and Controls

When you add a heat pump to your existing furnace, you will have to make some control changes. You will have to install a two-stage heating and cooling thermostat to control the heat pump. As part of the heat pump control circuit, there will be a terminal board with several lettered and numbered screws. Some of these terminals match the markings on the thermostat terminals, and similarly-marked terminals will be connected to install the thermostat.

From the terminal board, a fan relay will have to be wired into the W1 terminal or the heat terminal going to the heat pump. This will turn on the fan whenever the heat pump turns on. Without this fan relay, the fan will not come on when the heat pump is operating. The heat pump will not warm the heat chamber to the furnace fan turn-on temperature, and thus the fan will not come on. Chapter 4 discusses furnace fan switches. A cooling circuit fan relay will be connected in the cooling circuit at the terminal. This relay turns on the fan when the cooling compressor begins running. The wire from the second stage of the thermostat (W2) is connected to the gas

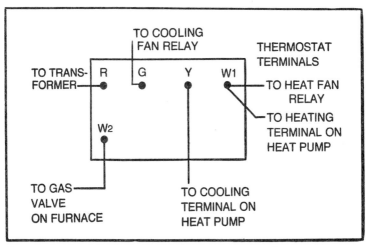

Fig. 6-8. This diagram of a heat pump two-stage heating/cooling thermostat shows where each terminal will be connected on the system. First, however, these terminals are connected to their corresponding numbers on the heat pump's terminal board, as shown in the next diagram.

valve, stack control heating terminal on the oil burner, or number one sequencer on an electric furnace, depending on what your furnace is. See Fig. 6-8 and 6-9.

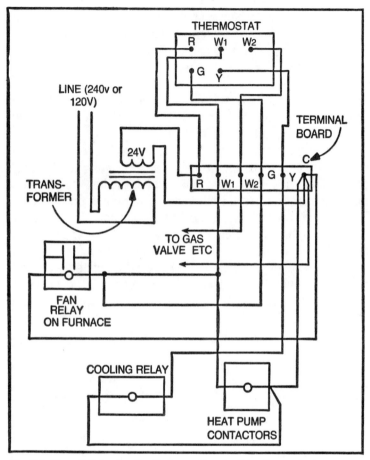

Fig. 6-9. The wires leaving the thermostat terminals go first to the heat pump terminal board, where corresponding thermostat terminals and terminal board terminals are connected, as shown. From there, the R terminal is connected to the 24 volt transformer. Incidentally, the line voltage coming into the transformer primary circuit will be 240 volts if the furnace is an electric furnace, 120 volts if it is a gas or oil furnace.

W1 connects to the heat pump contactors and the fan relay. With this connection, when the first stage of the thermostat comes on, the heat pump begins heating and the fan starts. W2 connects to the gas valve, the stack control or the first sequencer, depending on whether yours is a gas, oil or electric furnace. This is the second stage of the thermostat, which turns on the supplementary heat. G is connected to the fan relay to bring the fan on when the compressor starts on the cooling cycle. C is the common terminal to which all circuits connect.

Sometimes there will be one terminal board on your furnace and another terminal board with the heat pump control box. In this case, take the thermostat wires and connect them to the heat pump control terminal board, matching the like terminals. You will bypass the furnace terminal board.

Now, run a wire from the W2 heat pump terminal to the gas valve, oil burner stack control, or number one sequencer and back to C on the heat pump. Run another wire from W1 to the fan relay and back to the common terminal, run a wire from the G terminal on the heat pump to the fan relay. Remove the terminal board from the furnace and remove the wiring from the back of the terminal board since it is no longer needed. These are all the wries you will have to connect if the heat pump has its own terminal board because the heat pump contactors, the G fan relay terminal and the cooling relay are already wired internally into the terminal board.

After connecting the thermostat and the terminal board low-voltage wiring, you are ready to connect the power line to the transformer, which will be found in the heat pump control box.

Supplementary Heaters

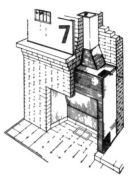

One way you may be able to save some money on your heating bill is by adding some supplementary heaters in often-used rooms of your house. You will have to do a fuel cost analysis as outlined in Chapter 3 to see if this will save you any money with your current fuel costs. Probably you will find that unless you are heating with an electric furnace now, you will save very little, if anything, by installing supplementary heaters.

Supplementary heater, as we are using the term, refers to any type of single-room heater that may be used either alone to carry the entire heat load of the room or to simply supplement the heat output of the central heating system in that room. Although supplementary heaters are available using a variety of fuels and designs, by far the most popular are electric resistance room heaters.

Supplementary room heaters can save you money if used properly because they give you an option with a central furnace of "zone temperature control." Zone temperature control simply means that you can adjust thermostats in different portions of the house to different temperatures, according to your needs. Thus, with supplementary heaters installed in some of the most used rooms of the house, you can turn the furnace thermostat down for the entire house. Set the room heater thermostats to bring the temperature up to a more comfortable level in the rooms you are using most often. Of course, you won't save any heating bill money if you turn down the furnace thermostat and then turn up room thermostats in all the house's rooms. But if your family can spend most of its winter evening time in one or two rooms and not be bothered by 60° temperatures in the rest of the house, you may save money by installing room heaters. Of course, you must consider the cost of the room heaters as an offset against any future savings. Generally, you can figure that you will save three percent on your heating bill or each 1° you lower the thermostat. But this saving is reduced by the amount of energy needed to supply the individual room heaters you then turn on. If you lower the thermostat for only a portion of the day, the savings will be correspondingly less than that three percent figure.

TYPES OF SUPPLEMENTARY HEATERS

Although electric resistance heaters are by far the most popular type of room heating units, there are other types available. Gas room heaters and electric hydronic individual room heaters are available. Among the electric resistance types are ceiling cable, baseboard heat and wall units. Ceiling cable is much like low-resistance electrical cable, except it has enough resistance to the flow of electricity to produce a substantial amount of heat for the occupants below. This cable in imbedded in the ceiling of a room. Pre-wired ceiling panels are also available. Baseboard heaters are resistance units that fit along the baseboard area of the room (Figs. 7-1 and 7-2). Generally these units rely totally on convection air currents and the principle that warm air rises to circulate the heat throughout the room. They do not usually have fans.

One type of heater that often has a fan to circulate the heat produced is the electric wall heater. This heater produces resistance heat, like the baseboard heater and ceiling cable, but it is attached to the wall of the room—usually about 2 feet above the floor. Some wall heaters do not have fans and depend on the natural convection air currents to distribute heat.

There are portable supplementary electric resistance heaters. These come with or without fans and plug into any 120 volt wall outlet. Usually they are too small to heat an entire room alone, although many of them do a very adequate job as supplemental units. Be sure to check the existing wire and make sure it is sized properly to take care of the heater to be

Fig. 7-1. The baseboard heater is a narrow elongated heater unit that installs at the base of the room wall. The length of the baseboard heater varies according to its heating capacity and design. The baseboard heater produces electric resistance heat, which is normally distributed by natural air currents without the help of a fan. You may connect the baseboard heaters so that each heater has its own thermostat or so that two or more heaters are controlled by the same thermostat.

used. There is an amp rating on the nameplate of the heater. If the circuit is almost overloaded with other appliances, some may have to be disconnected when the heater is used. Or you will have to use another circuit.

INSTALLING ELECTRIC ROOM HEATERS

This section discusses installation techniques for baseboard heaters, but the principles are certainly applicable to any room-sized electrical heater. Installing an electric wall heater would be about the same as a baseboard heater. Before you install any unit, however, read over the manufacturer's installation instructions included with the unit. There may be some specific installation directions for your heater.

Fig. 7-2. This is a baseboard heater with the cover removed to show the heater element underneath.

Selecting the Heater Size

The first step in installing a baseboard heater is to select the size heater you need. Generally, you can figure you will need about 1500 watts of electric heat capacity for each 100 square feet of floor area. Colder than average climates and poorly sealed and insulated homes will require a larger heating capacity. If you are using the electric heat for supplementary heat and not as the room's primary heat source, you can reduce the capacity needed to ½ or ⅓ that amount.

Baseboard heaters normally come in sizes from 500W to 2500W. Sometimes you will need to install more than one heater in a room. These may be installed end-to-end in a line, or they may be placed in different parts of the room. Usually it will be most convenient to operate two heaters in the same room from a single thermostat. Baseboard heater thermostats are discussed later in this chapter.

When you install two heaters on the same circuit, you must be careful the wire you use and the circuit breaker are large enough to carry the *sum* of the amperage ratings for the two heaters. Thus, if you install two 20 amp heaters on a single circuit, you will have to use at least number six or larger wire, and the circuit breaker must be 45 amps or so.

Installing the Baseboard Heater

Of course, the same thing holds true when installing a single baseboard heater—the wire and circuit breaker must be large enough to carry the current draw of the heater. On a 20 amp heater, for instance, this will mean at least number 12 wire (larger wire numbers mean smaller wire) or larger. If you are not sure of the amperage draw of your heater, you can find it easily by using the formula $I=P/E$, where I is the amps, P is the power in watts, and E is the voltage. A 1500 watt heater on a 240 volt circuit would draw 6.25 amps.

After you have run a line from your fuse box to the room where you will install the heater, you will connect the ther-

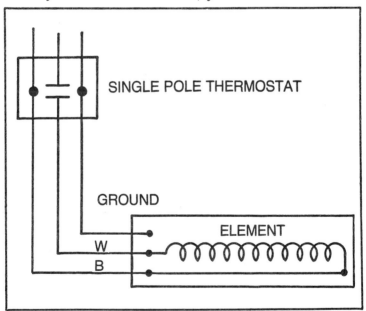

Fig. 7-3. This is the way most baseboard heater installations are wired. Usually just one heater is needed per room for supplementary heat. A three-wire electrical cable is connected to a single-pole thermostat. One wire is a ground wire that grounds the heater to the fuse box.

mostat to the power line, and then extend the line to the heater. Figure 7-3 shows a diagram of how a typical single baseboard unit is wired. Notice that grounded three-wire cable is used. This prevents a shock to room occupants who might touch the heater if it had a short circuit. Many baseboard heater units have their own thermostat wired internally and mounted on the heater case, so the installer only has to connect the power line to the heater terminals to wire in the unit. The built-in thermostat is already wired into the heater circuit for him.

Connecting Two Or More Heaters

Usually when you are installing baseboard heaters for supplementary heat, one heater per room will be sufficient. Occasionally, however, you need to install two baseboard heaters in the same room. This might be the case if you had added an additional room to your house and were depending on baseboard heaters for the primary heat source without extending the central heating system to that room.

You could install two separate heaters on two separate circuits—each with its own thermostat. But this wouldn't make much sense if the heaters were in the same room. Naturally, you want them connected to the same thermostat. Fig. 7-4 shows how these two heaters would be connected to a single 240 volt circuit. With two heaters connected to the same circuit, you will probably have to make the circuit a 240 volt circuit, to get the wattage you need and keep the amp draw of the circuit low enough.

When you have two heaters connected in the same circuit, the heaters must be wired in parallel. If they are wired in series, the heaters will get warm but will not reach a temperature high enough to heat the room.

If you wanted to connect two 120 volt heaters in the same room, you would not be able to connect them in the manner shown in Fig. 7-4 because the two heaters connected together would draw too much amperage at 120 volts. But you still could connect them through the same thermostat if you wired them as shown in Fig. 7-5. By using a double-pole thermostat, a single thermostat controls two entirely separately circuits. Thus, you could install two 120 volt heaters on the same thermostat.

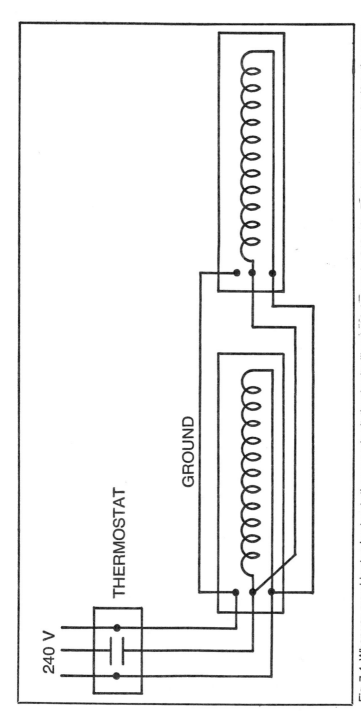

Fig. 7-4. When you are wiring two heaters to the same circuit, the circuit will probably have to be 240 volts. The heaters can be wired into the same single-pole thermostat, as here, but you must be sure the heaters are connected in parallel and not in series. This type of arrangement would work if you were installing two heaters in one room, or if you were installing heaters in different rooms but wanted them wired into the same thermostat.

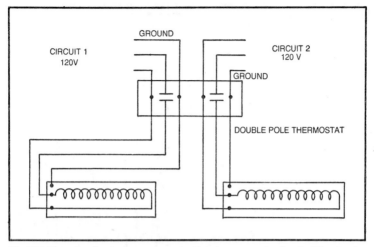

Fig. 7-5. Here, two heaters are connected into two separate circuits. The circuits are controlled by the same double-pole thermostat.

THERMOSTATS

As we have mentioned, some electric heaters come with their own thermostat already installed. On many, however, you will have to install the thermostat separately from the heater.

Room heaters almost always use line-voltage thermostats. See Chapter 11 for an explanation of the difference between line-voltage thermostats and low-voltage thermostats. The thermostat has an amperage rating that you should be aware of when you connect the thermostat. This rating tells the maximum current that should be sent through the thermostat. When you install a baseboard heater, you must be sure the amperage draw of the heater circuit is not larger than the amperage rating of the thermostat. For single heater installations, a 20 amp thermostat will usually be adequate; for two heaters through the same thermostat, you probably will have to go a 30 amp or larger thermostat.

You should install the baseboard heater thermostat with the same considerations in mind that you would have installing any thermostat. Chapter 11 outlines some of the do's and don'ts of thermostat installation and placement. For instance, you should be sure the thermostat is installed along an inside wall where it will receive good air circulation. It should not be installed along an outside wall. One of the disadvantages of having the thermostat built in as part of the heater unit is the thermostat cannot be separated from the heater. Therefore, you must try to install the heater unit along an inside wall if you have a heater with a built-in thermostat.

The line-voltage baseboard heater thermostat is usually installed the same way you would install an electrical outlet. A receptacle box is nailed to a wall stud, the thermostat wires are run into the box, and the thermostat body is installed on the box. Of course, this job is much easier on unfinished walls. If your walls are finished, you probably will want to consider purchasing a baseboard unit with a built-in thermostat.

BASEBOARD HEATER TROUBLESHOOTING AND TUNEUP

Properly installed, baseboard heaters will operate with just about as little maintenence as any heating units available.

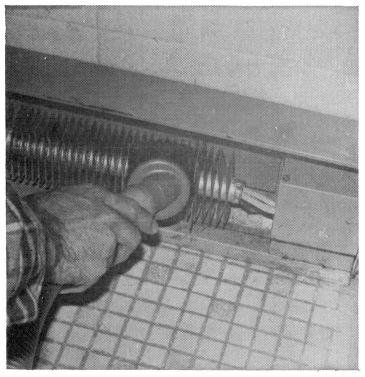

Fig. 7-6. Once a year you should give all baseboard and room heaters in your home a complete visual inspection to check for loose wiring connections. Remove the grill cover and vacuum out the heating element. Dust and dirt can collect on the element and fins, reducing the heater's efficiency.

As with any heating system, however, once a year you should give your room heater a thorough visual inspection and cleaning.

Check for any loose or damaged wire terminals. Remove the grill cover and dust off the grill facing. Most of these room heaters do not have filters, so they are susceptible to dirt and dust collection. Vacuum out the heating elements, as shown in Fig. 7-6. If your heater has a filter, clean it or replace it (see Chapter 4 for a discussion of air filters).

If your baseboard heater is not heating properly, it is a good idea to perform a voltage check on the unit. Turn the power on, and connect the voltmeter probes to the two power terminals. If you have a 240 volt heater that is wired into a 120 volt circuit, the heater will barely heat up.

If the heater will not heat properly, check for a loose connection, two heaters connected in series instead of parallel or a 120 volt circuit connected to 240 volt heater.

If the heater blows fuses, problem may be fuse too small for the circuit, a heater oversized for circuit, an element or a thermostat shorted out. If the heater will not heat at all watch for a loose connection, blown fuse, an element burned out or a thermostat malfunctioning. Short across the thermostat terminals to test it.

SUMMARY

Supplementary room heaters can save fuels costs if your family is willing to turn down the house thermostat considerably and allow the room heater to warm up the one or two rooms where you spend most of your time. You will save money.

If a fuel cost analysis (see Chapter 3) shows you will save more in fuel costs than it will cost you for the electricity for the heaters, you will save money. Usually only those families using electric resistance heat will save enough to warrant the room heater installation. The cost of purchasing the heaters must be counted as an offset against any future savings.

Once a year, give your room heaters a visual inspection and cleaning.

The Duct System: The Final Link

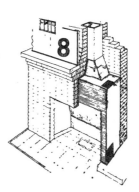

The duct system of your furnace is the last leg of your heating system. It is the final link that takes the warm air produced in the furnace to the living area where it will do some good. But just because the duct system is the final link in the complete furnace system does not mean that it is an unimportant one. Unfortunately, in many home heating installations the duct system appears to take a far back seat to the rest of the system. The duct system, seldom seen and difficult to reach, is simply ignored by most homeowners.

But this should not be. In many heating systems, the duct system accounts for a heat loss of 15 percent or more. If you could recover only half of that, it would be a substantial savings, especially at today's prices.

This chapter is devoted to explaining how you can improve your furnace's duct system to cut wasted heat energy and fuel dollars. We show you how to inspect your duct system and how to make duct modifications when necessary.

DUCT SYSTEM BASICS

The basic parts of the duct system are these, in the order the furnace air flows through them: plenum chamber, main duct, run and room register. There are a number of additional connecting joints you can purchase or make to connect these primary parts together. Figure 8-1 shows some of the major duct system components.

Most duct system parts, from the round straight pipes used for the runs to the boots that deliver the furnace air to the rooms, are available in commercially prepared units that can be easily connected and fitted together. Few heating installers make their own ductwork anymore from plain sheetmetal. You can buy these factory-made duct parts from most heating supply shops or from sheet metal shops, and you will find them much easier to work with—especially as a beginner—than making your own parts. This is particularly true for connecting parts such as take-offs, boots, reducers and plenum connectors. If you are considering installing some ductwork and are not sure of the parts you need, take a diagram of what you have in mind to the heating system supply house. Usually they can help you a great deal in getting exactly the duct parts you need.

Most ductwork is purchased with plain sheetmetal walls, but you can purchase duct parts in pre-insulated styles. This type comes with sheet metal exterior walls with a glued-on insulation interior. Of course, these pre-insulated parts are

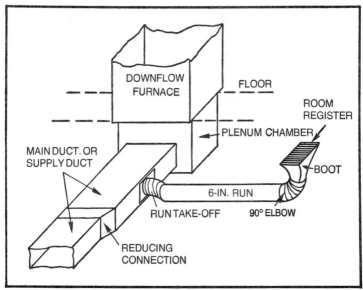

Fig. 8-1. Here are some of the main parts of the duct system. A number of commercially prepared duct pieces are available, which make duct installation much simpler than making the pieces yourself.

Fig. 8-2. A take-off connection.

usually more expensive than the noninsulated ones plus the cost of insulation. So you can usually save money by insulating the ductwork yourself.

The plenum chamber is a part of the duct system that is actually an extension of the furnace. On an upflow furnace, the plenum chamber extends above the furnace. On a downflow furnace, the plenum chamber extends below the furnace and under the floor. From the plenum chamber, the furnace air goes into one or more main ducts.

The main duct, also called the supply duct or the main trunk line, is a large duct that extends from the plenum chamber and goes the length of the duct system supplying air to the individual room *runs*. If the furnace is in the center of the house, there may be two main ducts, one in each direction from the plenum chamber. The main ducts may be constructed on either rectangular or round ductwork, but rectangular is much more popular because it requires less vertical height. A main duct connector or a plenum take-off attaches the main duct to the plenum chamber.

The main duct itself does not take the furnace air into the individual rooms. That is the job of the *runs*. The runs are normally constructed of round 6-inch or larger duct pipe. These are connected to the main duct by means of "take-offs." You can make your own take-offs from round duct pipe, but it is much easier to use commercially prepared take-off connections (Fig. 8-2). When purchasing take-offs, you must be sure to get a size that will fit the run ducts and the main ducts you plan to use. At the end of each run is a boot that empties the air out into the living area. Usually you will also have to have an elbow connection to connect the run with the boot.

The room register directs the flow of air out into the room. A number of different types of room registers are available. Some have adjustable louvers, and some have vents which close.

The air flow through the duct system is maintained by constant static pressure in the system. To maintain this pressure, the size of the main duct must be reduced. Commercially prepared reducing connections are available to do this easily. Without reducing the duct, you would find that the rooms served by runs at the end of the main duct would receive very little warm air.

With commercially prepared ductwork available, piecing together an entire duct system has become a relatively simple job (compared to times past, that is). But there is more to it than just fitting all the pieces together. For one thing, every connection must be fastened together with screws or rivets. A ¼-inch drill with a bit about the size of a large nail should be used to drill at least two holes in every connection—one on each side—to hold all pieces together. Insert sheet metal screws into the holes, or use the quicker and easier pop-rivet gun, as described in the next section. You should fasten all parts together as you connect them.

The larger parts of the duct system must be suspended from the floor joists above unless you are installing a duct system in the attic that will rest on the ceiling joists. You can suspend the ductwork using wires or 1-inch strips or sheet metal nailed into the floor joists. See Fig. 8-3.

The duct system should be insulated if it runs through an unheated area such as an uninsulated crawl space or basement. Insulating the duct system will cut heat loss through the ducts, and insulation is imperative if you have central air conditioning. Commercial duct insulation is available to insulate the ducts, but you can use 2-inch or thicker fiberglass batt or roll insulation. The insulation should have a vapor-barrier backing. You will find it is easier to insulate each large section

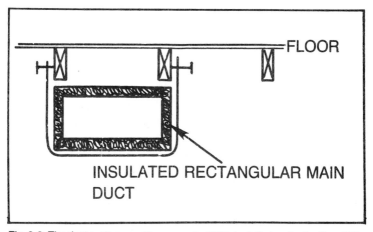

Fig. 8-3. The duct system must be supported if it is installed under the floor. This support can be most easily provided by hanging the duct in the manner shown. A wire or 1-inch strip of sheet metal nailed to the floor joists to suspend the duct works well.

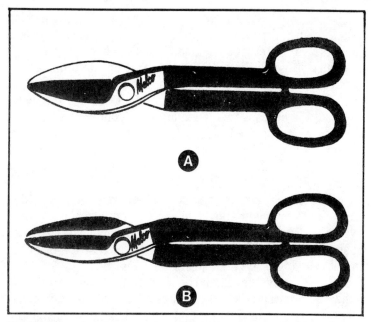

Fig. 8-4. These tin snips are a scissors style, and they are available in regular pattern (A) for straight line cuts and combination pattern (B) for circular cuts and straight line cuts. This style of tin snip requires you to separate the cutting jaws manually, thus making cutting more difficult than with spring-loaded snips. You can purchase these tin snips at most hardware stores.

of ductwork before installing the ducts. Insulating the ducts is discussed later in this chapter.

DUCT TOOLS

The one tool you will have to have if you do any duct work at all is a pair of tin snips. There are several types of tin snips available, and depending on how much duct work you do, you may want to purchase several types or just one type. If you are going to purchase just one type of tin snip, we recommend you buy what is called a *combination pattern* pair of tin snips. These snips are designed to cut either a straight line or a circular cut. A *regular pattern* pair of tin snips can be used primarily to cut straight lines. Regular pattern snips and combination pattern snips are shown in Fig. 8-4.

Another set of tin snips is shown in Fig. 8-5. This style shorter handles that are spring-loaded for easier cutting, and the blades are twisted to make circular cutting easier. You can purchase these tin snips in right-side circular cut, left-side circular cut and straight line cut patterns. If you do quite a bit of ductwork, you probably should purchase a pair of center-cut snips. Not to be confused with straight-line snips, the center cut snips cut on two edges at once so that a thin ribbon of sheet metal is removed as you cut along a line. These snips are excellent for cutting a circular hole or for cutting in a long straight line. With center snips, you do not have to keep bending the cut metal out of your way.

If you are doing only a modest amount of duct work, a screwdriver and sheet metal screws will suffice as a fastening system. But if you install very much ductwork, you will want to make the job easier with a poprivet gun, like the one shown in Fig. 8-6. Using this hand-held device, you can place a rivet into the hole you drill to fasten the joints of duct together. It will require only a fraction of the time necessary to insert a screw into the drilled hole.

SIMPLE DUCTWORK AND ENERGY SAVINGS

Most of the energy-saving ideas in this book require no ductwork at all, but a few require a small bit of basic ductwork. Normally, those cases requiring ductwork are very simple, such as connecting a wood stove into your heating system.

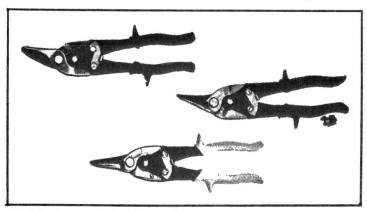

Fig. 8-5. These spring-loaded type snips are easier to use than the standard scissors snips because the cutting jaws open under spring power. These snips are available in right-circular cut, left-circular cut, straight cut and center cut models. You can purchase them at most hardware stores or heating supply stores. The twisted jaw design of these snips makes them easier to use and to cut circular patterns than the scissors type of snip.

Assume you wanted to connect a fairly large circulator wood heater into your present furnace system, as described in Chapter 9. What you want to do, as shown in Fig. 8-7, is to run a duct between the output connection on the wood stove and the cold air return on the existing furnace. This is a fairly simple change in your duct system, but it is one that can save you quite a bit of money.

You will first have to have the stove manufacturer's specifications on the size of duct pipe needed. You need to examine your furnace room to determine where you will place the stove in relation to the furnace; then determine what the connecting duct line will look like. You need to determine the approximate length of the joints you need and the number of turns the duct will make before connecting with the cold air return. In our diagram, we need two straight joints and one 90° elbow. You need a take-off connection to fasten the round pipe into the cold air plenum.

Once you have the duct parts, begin by marking the size of the hole you will have to cut in the cold air plenum to install the take-off. Cut out this hole and install the take-off connection. Fasten it with screws or pop rivets. Then, simply add each joint of duct pipe and fasten it until the duct is completed. In some cases, you may have to rig up some sort of suspension system to support long lengths of ductwork. As your final step, you should inspect for leaks and insulate the duct.

DUCT REGISTERS AND THEIR PLACEMENT

Air distribution registers may be placed in several different locations and they are available in several different designs, but you usually will find only one or two types in any given home. The better registers for purposes of energy efficiency are those that have dampers which close to adjust the flow of air through the register. This lets you cut down the amount of heat that reaches some rooms of the house if you decide you do not need so much heat in that room. Some registers also have adjustable louvers that allow you to direct the air flow. Sometimes this can be a helpful feature on a duct register, but you can expect to pay extra for it.

It is important to note that if you have central air conditioning you will need different duct registers. Heating/cooling registers have more space for air flow because of the difficulty

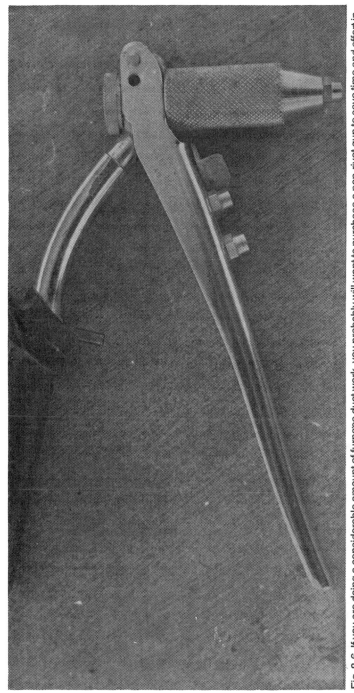

Fig. 8-6. If you are doing a considerable amount of furnace duct work, you probably will want to purchase a pop-rivet gun to save time and effort in fastening the joints of duct together. You will still have to drill a hole in the duct joint walls to fasten them, but you can insert a pop-rivet in that hole to fasten the joints in much less time than you can insert a sheet metal screw.

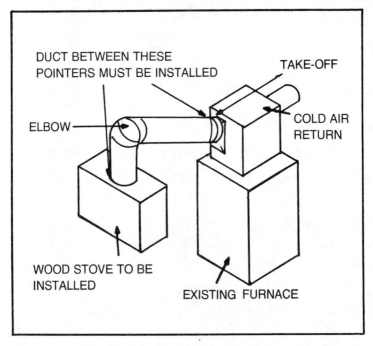

Fig. 8-7. To install a wood stove as an addition to your furnace system, and to connect the wood stove into the duct system, you will have to run a duct from the wood stove outlet to the cold air return plenum on your existing furnace. Chapter 9 explains how you can select a wood stove or small furnace for this use. The first step in installing the duct is to determine where the stove will sit in relation to your furnace. You must determine about how many of what length and size duct you need. You also need to determine how many elbows you will use. You will need a take-off to attach the round duct pipe into the cold air return.

in moving cool air. Thus, during the heating season, if you have a central air conditioning system, you might want to close the air flow damper on your registers a bit.

Figure 8-8 shows how some common types of registers would be placed in a room. Generally, floor registers and baseboard registers are placed along an outside wall and near any windows. This puts the registers where the largest heat loss is, and it also tends to draw the cooler air from around the windows into the return air duct stream where it can be heated by the furnace. Baseboard and floor registers can certainly be placed along an inside wall if needed. The baseboard registers fit right against the base of the wall, where they connect to the boot. The floor registers are set into the floor an inch or so

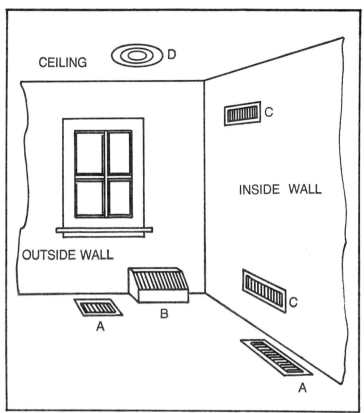

Fig. 8-8. The registers may be placed in a number of locations throughout the room. Registers are constructed for different applications. Floor registers (A) are placed into the floor, several inches away from the wall. If possible, they should be placed along an outside wall where the largest heat loss is. They may, however, be placed on an inside wall. Baseboard registers (B) fit flush with the wall at the wall's base. They too should be located along an outside wall when possible. Wall registers (C) may be located either high in the wall or low in the wall, but they must be placed on an inside wall. This is because the duct that supplies them is run behind the wall covering. If that duct were run in an outside wall, there would be room for almost no insulation between the duct and the cold outside air, and this would create a large heat loss in the duct. Ceiling registers (D) are usually circular registers, but they may also be flat rectangular designs.

from the wall, and they are placed parallel to the wall (Figs. 8-9A and 8-9B).

You should not place a wall register in an outside wall. The wall register connects with a duct that runs partway in the wall, and there would be a lot of heat loss if this duct ran through an outside wall where there could be little if any insulation between the duct and the cold outside air. There-

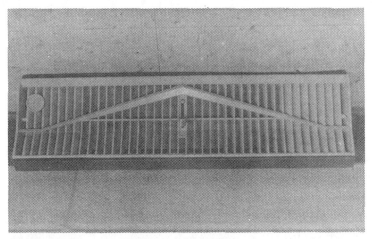

Fig. 8-9A. Here is a baseboard register that installs upright against the wall at the base. The pointer indicates the air flow adjustment lever.

fore, a wall register is always placed in an inside wall. The wall register may be a high wall register or a low wall register, depending on where in the wall the register is placed. A high wall register will be used if the duct system is above the room in the attic, and a low wall register will be used if the duct system is beneath the floor.

The ceiling register is placed in the ceiling, and it is used when the duct system runs above the ceiling. The advantage of the ceiling register is that it distributes air in a circular fashion all about the room. Its major disadvantage is that during the heating season most of the warm air from the furnace may stay at the top of the room, since warm air rises. When the heat register is placed in the floor or low wall, the air will circulate from the floor to the ceiling, where the return air duct is located.

You can lose a lot of effective heat from your furnace if you are careless about register placement. The registers should not be placed behind draperies, large plants or furniture. To get the heat from the furnace out into the room where it can warm the occupants, there must be a free air path in front of the register. You may want to use a floor register placed several inches in front of a window to keep it away from floor-length draperies. You would not want to use a baseboard register where the drapes would block the airflow.

ADJUSTING THE REGISTERS

For the most efficient air distribution, you should have adjustable registers so that you can alter the air flow as needed. If you really look carefully at the heating needs of your home, you will discover that there is no reason for all registers to be wide open. Sometimes you may want to entirely close off a room. Naturally, this will save you some money on your heating bill. Also, the different rooms in your home have different heat requirements. The kitchen, for instance, has a number of appliances that generate heat, so you probably can get by with closing the kitchen register a bit. If you have a room with a fireplace or a wood stove, you should close the heat register when you have built a fire. South-facing rooms with large window areas normally require less heat, while north-facing rooms require more. In some little-used rooms, you may want to cut the air supply in half. Anytime you can, close a register. Even if it's only a little bit, you'll be saving fuel money.

A good way to adjust your registers is to use a thermometer to check the temperature of one or two rooms. The best time to do this is at night when the sunlight will not affect the temperature reading. If the temperature in the room is higher than your thermostat setting, close the registers a bit. About a

Fig. 8-9B. A floor register that installs flush with the floor an inch or two away from the wall.

day later, check the room termperature once again. By adjusting the registers throughout the house a room or two at a time, you can achieve a desirable and economical air flow adjustment.

INSULATING THE DUCT SYSTEM

When your duct system runs through an unheated area such as a crawl space or an attic, it can really lose a lot of the heat the furnace has produced. Bare sheet metal walls are excellent conductors of heat, so heat transfers easily from the ducts to the cold air surrounding the ducts. That heat is heat you are paying for but have lost.

Insulating the duct system will prevent most of the heat loss in the duct system. Plus, there are added benefits. The insulation you install should have a vapor barrier backing (such as reflective foil). This vapor barrier prevents moisture in the crawl space from rusting out your duct system. Thus, your duct system will last longer.

If your duct system is under your house in the crawl space, your first step in reducing duct heat loss is to seal any foundation vents. These vents are necessary to evaporate moisture under the house, but they needn't be open during the winter. The drafts they allow in the crawl space will take a lot of heat out of the ducts, and they will also make the floor of the house much cooler. You also should seal any other foundation holes, such as those that water pipes pass through.

Of course, the easiest time to insulate the duct system is when it is first installed. You can buy ducts that have insulation material already applied on the inside of the metal exterior. Usually you can save money by purchasing the duct and your own 2-inch or thicker foil-backed fiberglass batt or roll insulation. Wrap the duct parts as they are installed. You will need lots of duct tape—2-inch wide cloth tape—to secure the insulation and seal all the seams. Figure 8-10 shows an insulated duct run.

When a duct system is run through a basement area, it often is not insulated. The theory behind this is that any lost heat is retained in the basement, and a warm basement is desirable. The problem with this thinking is that unless the basement walls are insulated with 3 inches or more of fiberglass batting insulation, they will lose quite a bit of heat

Fig. 8-10. When your duct system runs through an unheated crawl space, basement, attic or garage, you should insulate the ducts to prevent heat loss. Fiberglass blanket insulation with vapor barrier backing works very well for this purpose. You should secure the insulation to the duct with furnace duct tape. Be sure to seal all seams in the insulation with tape to prevent heat leaks.

into the surrounding earth. The heat you allow to escape from the ducts into the basement then becomes rather expensive. Therefore, unless your basement walls are insulated, you really should insulate the duct system running through it just as you would if the ducts ran through an unheated crawl space.

If your perimenter foundation walls are insulated with fiberglass insulation or with foam insulation, you will not need to insulate the duct system. When the foundation walls are well insulated and sealed, we say the homeowner has a *heated crawl space*. Although it is true that you will lose some heat from your duct system in a heated crawl space, by insulating the ducts you probably will not save enough extra heat to pay for the cost of insulating them.

If your duct system is not insulated and your crawl space is not insulated, you are losing quite a bit of heating fuel money. To insulate a duct system that is already installed, you can wrap blanket insulation around the ducts, or you can nail insulation to the floor joists on each side of the duct run so that the run is insulated. See Fig. 8-11.

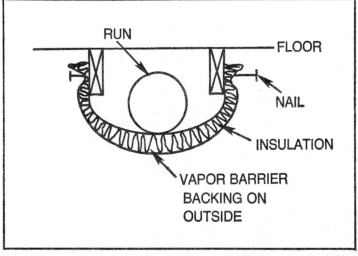

Fig. 8-11. When your duct system has been installed without insulation, you should insulate it to achieve substantial savings on your fuel bill. If you cannot wrap the insulation around your ducts, you can still insulate the ducts by nailing fiberglass blanket insulation to the floor joists on each side of the runs. Use 2-inch or thicker fiberglass batting insulation with a vapor barrier backing. The backing should face the crawl space. Be sure you attach the insulation well enough to the floor joists to form a seal along the edges of the insulation. This will prevent heat leaks.

SUMMARY

1. The duct system should be insulated to cut heat loss if it runs through an unheated attic, basement, crawl space or garage. All insulation seams should be well sealed to prevent air leaks.

2. Cracks and foundation vents should be closed when the duct system runs through the crawl space.

3. For energy efficiency, the better room registers have vents which close, allowing you to adjust air flow.

4. Registers should be placed where they are free from air flow obstructions such as drapes and furniture.

5. The air flow through the registers should be adjusted for each room to meet its heating needs. Turning down the air flow into a room saves money.

Saving Money With Wood and Coal Heat

Since the oil embargo in 1973, more and more attention has been paid to wood and coal as a source of residential heat. Wood heating especially has become quite popular as people began looking for alternative, less expensive methods of heating their homes. Not the least of the reasons for increased attention and interest in wood heating has been the large media coverage of wood heating methods and promotional campaigns of firms that stand to profit handsomely from increased use of wood heating.

The booming popularity of wood heating brings America's heating fuel demands almost a full circle from the early days of the nation when wood supplied almost all the heating fuel for a growing country. Of course, there are large differences between wood heating now and then. Decades ago, homes were smaller and wood was in plentiful supply in most parts of the country, so an efficient wood heating system was not as important as it is today. Today's homeowner who wants to burn wood must consider where he will obtain his supply of wood, its cost, and the cost and availability of modern heating fuels such as oil, electricity and gas.

If you are considering installing a wood or coal heating system primarily for its heating value—as opposed to the aesthetic value of such fuels—you are probably motivated by two factors: lower heating costs, and a steady, reliable supply of heating fuel that will be available no matter what type of federal energy controls or foreign policy are in force. In planning your wood or coal heating system, you should be sure to keep these two goals in mind. Otherwise, you may find yourself in a position you sought to avoid—having to rely upon a steady supply of gas or oil to operate your wood furnace.

It is possible to install a heating system that relies entirely on wood or coal. We know of a number of homes that have never been heated with any other fuels! The bulk of this chapter, however, deals with using wood or coal as a supplementary source of heat where the wood or coal units are used in conjunction with an existing gas, electric or oil furnace. Using this type of system, wood or coal can become your primary heating fuels and the "modern" fuels can be used as back-up fuels for those times when the wood or coal supply gets low or the wood or coal burning unit is inadequate to heat your entire home. If you intend to make coal or wood your only heating fuel in a central heating system, you will find these systems discussed later in the chapter in the section dealing with wood and coal furnaces.

Because coal and wood are essentially interchangeable as fuels, this chapter discusses coal-burning and wood-burning units together. Many stoves and furnaces can be easily adapted to burn either type of fuel.

CAN YOU SAVE MONEY?

For many persons, the amount of savings is not an important consideration in their decision to install a wood-burning fireplace or stove because they find an occasional wood fire aesthetically pleasing. But if your primary interest in burning wood or coal is to lower your fuel bills, determining whether you will save money decides whether you will install such a system at all.

Figuring if you will save money by installing a wood or coal burning unit has become more difficult in the last few years because of the claims and promotions made by companies selling these units. Many claims of potential savings are drastically beyond those a homeowner could reasonably expect to achieve. They are inflated advertising claims designed to take advantage of the growing wood heating market and increased concern over higher fuel prices. To find out if you save money by using wood or coal as a heating fuel, you must do an analysis of fuel efficiency and cost similar to that outlined in Chapter 3.

Table 9-1 shows how wood and coal compare with other common fuels in heat potential. You can generally expect that one cord of hardwood holds about the same BTU potential as one ton of coal. (A "cord" is a tightly packed stack of wood $4' \times 4' \times 8'$). Softwood will require almost two cords of wood to get the same heat potential contained in one ton of coal.

But Table 9-1 shows the heat *produced* by wood and coal compared to other fuels. It does not account for variances in efficiencies of the furnaces and stoves in which these fuels are burned. Since a wood stove is only about half as efficient as a gas or oil furnace, the wood must produce twice as much heat per dollar just to break even in heating fuel cost with your present system. The efficiency of wood and coal-burning units is difficult to pinpoint with any degree of accuracy because efficiency varies greatly with the unit and installation. As a rule of thumb, you can figure that a thermostatically controlled wood stove will be about half as efficient as a gas or oil furnace and about 40 percent as efficient as an electric furnace. Therefore, for a wood stove to save you any money at all on your heating bill, wood must cost less than half as much for the amount of heat produced as gas or oil. Since a fireplace is even less efficient than a wood stove, your wood must cost less than one-third as much as gas or oil for the amount of heat produced —assuming you have a relatively efficient fireplace.

Thus, as an example, if you are thinking of installing a heat-circulating fireplace, one cord of maple wood must cost less than one-third as much as 135 gallons of heating oil in

Table 9-1. This chart shows the amount of each fuel needed to equal the heating value in one cord at the specified wood.

Type of Wood	GAS, CU. FT	OIL, GAL.	COAL, TONS	ELECT., KWH
HARDWOOD (MAPLE) AND COAL	20,000	135	1	5500
HARDWOOD (OAK)	24,000	165	1.25	6700
SOFTWOOD	14,500	95	.65	3900

order for you to save any money on heating bills by adding the fireplace. As you can see, you may not save much money on your fuel bill by burning wood or coal unless you can obtain a relatively cheap supply—something that is becoming more and more difficult lately. For a more complete discussion of how to determine fuel cost factors, see Chapter 3.

Obtaining a Supply of Wood or Coal

Since one reason for the growing popularity of wood and coal heat is the desire to insure an adequate supply of heating fuel, you should carefully consider your likely sources of fuel before deciding to install a wood or coal heating unit. Coal at one time was plentiful in almost every city, but its decline as a residential fuel has made it difficult to obtain in many areas. Before deciding to install coal-burning units, check sources of supply. See if the supplier will deliver to your home or if you will have to arrange for delivery yourself. What quantities will you have to purchase as minimums?

Many persons have been recently using wood heat because they have access to free or very cheap supplies of firewood. Of course, they generally must cut, haul, and stack the wood themselves. But if you live in an urban area and must purchase wood at the prevailing market rate, firewood will not be cheap. Firewood costs have exceeded $100 a cord in many urban centers during recent winters! At those prices, it is very difficult to save any money heating with wood. Consider the supply of standing timber in your region. As the demand for wood increases, prices will rise unless there is a large supply of potential firewood available. If you are purchasing firewood, you can expect prices to increase in coming years, but these increases that would cut into potential heating fuel savings may be offset by expected increases of prices for other heating fuels.

Storing Wood and Coal

Storing wood and coal may be the single largest handicap of these fuels. Storage consumes a good deal of space, and it can be messy. Wood must be kept dry to keep from wasting half its heat energy by evaporating the moisture it will absorb—a waste you can ill afford if you purchase your wood. Check your local building code, zoning ordinances and subdivi-

Fig. 9-1. This is a modern-style circulating stove. The outer stove surface visible here is a sheet metal jacket that encircles the firebox. Air circulates between the jacket and firebox walls, where it is warmed before returning to the room. Stoves of this type are covered with a porcelain surface for easy cleaning, and they are available in a wide range of colors (courtesy Shenandoah Manufacturing Co.).

sion restrictions to be sure there are no rules that would prohibit you from storing your fuel outdoors. If there are, you may have to arrange some way to store it in a garage area or in a separate building near the house.

Building Codes and Insurance

Almost all cities have building codes that prescribe strict safety precautions that must be followed in installing a wood heater, fireplace or wood furnace. Check with your local city hall and fire department and obtain a copy of these restrictions before you finalize plans to install a wood-burning unit. Usually, adding a wood-burning unit will not affect your home fire insurance rates if the unit is properly installed. You must report the installation, however, or your fire insurance policy may not cover fires caused by the wood-burning unit.

If your home has a mortgage, you also may have to report the installation of a wood-burning unit to the mortgage holder. As long as proper installation procedures are followed and necessary insurance is obtained, you should have no problems in getting the approval of the mortgage holder.

WOOD AND COAL STOVES

Generally, wood-burning heating units are classified into three types of systems. The *fireplace* has an open flame, visible to the occupants of the room being heated. The *stove* is a metal enclosure in which the wood burns. The flame is contained inside the enclosure and is not visible to the occupants of the room being heated. Stoves may be used to heat more than one room and may even include a blower fan to circulate the heat around the house. But they are not designed to operate as part of a central heating furnace system with ducts and registers in each room. The *furnace* is designed to serve as a central heating unit, located out of the living area and connected to the rooms with a duct system. Coal-burning units may be either furnaces or stoves, but coal is not used in fireplaces.

Stoves have several practical advantages. Most modern models are reasonably efficient. For the amount of heat they produce, they generally are inexpensive. And they are available in a wide variety of styles and sizes, so a homeowner can install a small stove as a one-room supplementary heater. He can install a large stove that will heat several rooms.

TYPES OF STOVES

There is a wide range of styles and designs of wood stoves available today, and you can see many of these designs by visiting a large hardware store, department store or stove supplier. You can purchase a traditional potbelly stove that must be adjusted manually and that will heat one large room or two small rooms. You can also purchase a modern stove that will heat five to seven rooms and comes equipped with a

Fig. 9-2. This is a more traditional-style radiant stove. The steel outer wall is actually the firebox, and most of this stove's heat is produced by radiation. This stove model has a thermostatic damper control, but many similar stoves must be controlled manually (courtesy Shenandoah Manufacturing Company).

thermostatically controlled damper, an attractive and easy-to-clean porcelain finish, a large firebox that holds up to a 12-hour supply of logs, and even an electric blower unit to help distribute the warm air through the house. Prices vary, but you can expect to pay about $150 for the manually controlled stove, and $350 to $500 for the thermostatically controlled modern-style stove, with models available at all prices in between.

Two styles of stoves are shown in Figs. 9-1 and 9-2. These stoves are just two of the hundreds of styles and designs available. With a little shopping around, you can locate a stove that fits your needs and budget.

Most of the stoves described in this chapter are available in models that will burn coal. However, if you plan to use coal instead of wood, you should check with your stove dealer to be sure he understands your needs and is selling you a coal-burning model. Often the only difference between the wood-burning and coal-burning models is the grate, but some stoves are designed to accept one fuel only.

The major recent changes in stoves have been in the area of improved efficiency. The traditional stove required manual adjustment of the damper—a very inefficient system relying entirely on guesswork by the operator. A manually controlled stove has an efficiency of about 20 percent. Merely adding a thermostatically controlled damper will increase the stove's efficiency quite a bit, which puts more heat into the house and less heat up the flue. Bimetal thermostats that automatically adjust the damper to control the rate of burning are now available on most stoves are well worth the additional cost. See Fig. 9-3.

Other recent developments in stove design have improved the burning process so that logs burn slowly and more completely to eliminate wasted heat going up the flue. Some modern stoves have a secondary burning process that burns gases produced by the wood and recovers some of the heat present in the exhausted air that goes up the flue. See Fig. 9-4.

BUYING A STOVE

If you are buying a stove to save money, the unit's heating efficiency should be your largest consideration in de-

Fig. 9-3. A cutaway view of a radiant stove. (A) is a bimetallic thermostat that opens and closes the damper as the heat expands and contracts the thermostat metal. (B) represents firebricks that line the firebox. (C) is the grate (courtesy Shenandoah Manufacturing Company).

ciding what type of stove to purchase. While a manually operated potbellied stove may be nostalgic and have visual appeal

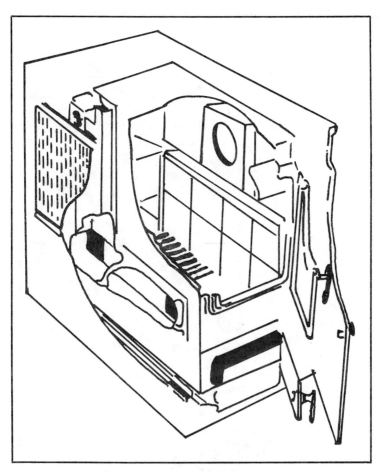

Fig. 9-4. Cutaway view of a modern-style wood and coal stove.

in some settings, the fact is that as much as 80 percent of the heat produced by the wood may be lost up the flue. That means that for each 10,000 BTU burned in the stove, a paltry 2,000 BTU make it into the living area! A more modern (but also more expensive) stove equipped with a blower and thermostat will be about 50 percent efficient and will put 5,000 or more BTU into the living area. That's a big difference if you do not get your firewood free!

You also want to consider buying a stove that has a large capacity firebox. The larger the firebox, the less often you'll have tend the stove and fill it with fuel. Many stoves today will hold a 12-hour supply of wood or coal, which is easily enough to burn the entire night and leave a hot bed of coals to start a new fire in the morning.

If you plan to heat more than one room with your stove, a blower system is a must. The blower removes much more of the heat from the stove and helps distribute the heat through the house. Without the blower the only air circulation will be through natural air currents, and not as much heat will be transferred to the air.

You can buy a stove in almost any design and color to match the decor of any room of the house, so a stove can add to the attractiveness of a room and need not appear out-of-place.

When you buy a stove, examine it with an eye toward what you will think of it after a year's use. Are the ashes easy to remove? Is the firebox accessible? What size logs will it accept? Remember that the larger the firebox opening and firebox length, the larger the logs you can put in the stove. It's

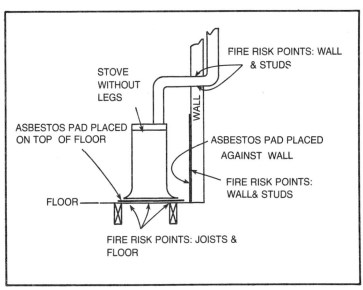

Fig. 9-5. A stove improperly installed can create a fire hazard even though the wall and floor are protected with noncombustible materials. Traditional-type stoves that have an uncovered firebox surface in the living area are particularly susceptible to these dangers. The heat will radiate from the stove through the noncombustible material and will ignite the wood on the back side. These dangers can be easily avoided by following manufacturer's instructions on minimum spacing between the stove and combustible materials and by providing an air circulation space between the wall and noncombustible protective material.

a lot more work to saw and split wood into small chunks, and that means more work for you or more money you'll have to pay for smaller wood if you buy a stove that accepts only small logs.

Check the inside of the stove with a flashlight to be sure there are no cracks or weak joints. Examine the entire stove for good craftsmanship strong welds and tightly fitted joints. The stove has to withstand intense heat, and a poorly constructed stove will warp, crack and break.

INSTALLING A STOVE

From the standpoint of safety, the most important part of adding a stove to your home is proper installation. If the stove is not installed properly, you run a large risk of fire in structural portions of the house that are not even visible from the living area. See Fig. 9-5. Even though you have taken precautions by placing noncombustible materials between the stove and the combustible wall and floor materials, wood structural materials can begin smoldering behind the noncombustible materials if the stove is placed too close to the wall or floor. To prevent these problems, provide adequate spacing between the stove and all combustible materials—even if they are covered with a protective noncombustible material. If you want to place the stove closer than 2 feet to a combustible wall, you may be able to do so according to manufacturer's instructions, providing you allow a 1 inch spacing between the wall and noncombustible material. (Fig. 9-6).

Before you install a stove, always obtain three things: a list of local building code restrictions on stove installation, a list of local fire code restrictions on stove installation, and owner's manual telling you how to install and operate your stove. All three should be easy to obtain. You should be able to get a copy of the installation instructions for your stove when you purchase it new. If you purchase a used stove, look for the stove's make and model number and write to the manufacturer requesting a copy of installation instructions.

Just about any do-it-yourselfer can install a stove in his home using simple tools and equipment. You must follow the manufacturer's guidelines, however, to avoid safety risks. Following is a general summary of installation safety guidelines, but always follow your stove manufacturer's re-

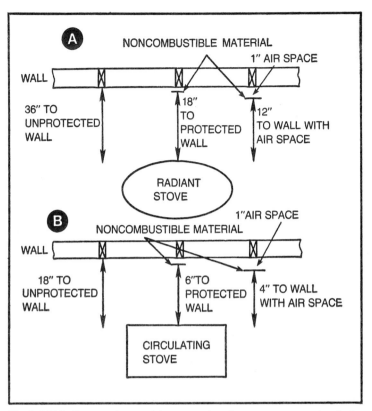

Fig. 9-6. This diagram shows minimum spacings between stoves and walls for six different stove and wall applications. (A) shows minimum spacings for noncirculating stoves. A noncirculating stove is a traditional-type stove that has no outer jacket enclosing the firebox. Thus, heat from the stove enters the living area by radiation or by direct contact with the sides of the stove. If the wall is unprotected, leave at least 36 inches clearance between the stove and wall. If a noncombustible material, such as asbestos millboard attached to sheet metal protects the wall, the stove may be moved closer to the wall. (B) shows minimum spacings for circulating stoves. Circulating stoves have an outer jacket that encloses the firebox so there is an enclosed air space between the living area and the hot metal walls that hold the fire. Room air clutches through this air space and is heated. Because this jacket enclosure retains much of the stove's radiant heat, a circulating stove may be placed quite a bit closer to a combustible wall than a noncirculating stove.

commendations when further safety precautions are recommended.

Protecting the Floors and Walls

As a general rule, a stove that has no blower system should be at least 3 feet from an unprotected wall constructed of combustible materials. In modern homes, this restriction applies to almost all walls, since even many "noncombustible" wall materials (such as gypsum wallboard) still have a combustible surface or covering, such as paper. Also keep in mind that the intense heat generated by the stove can ignite the wood studs covered by the wallboard if the stove is placed too close.

If the stove has a blower unit to remove the heat from the unit, it may be placed a bit closer to an unprotected wall. The *circulating* stoves have an outer jacket around the firebox, and this jacket contains radiant heat and protects the wall to a large extent. These stoves may be placed within about 18 inches of an unprotected wall, but check the manufacturer's instructions to be sure. Never place your stove closer to a wall than the manufacturer recommends.

You may place your stove closer to the wall if you provide some type of wall protection made of a noncombustible material. A good wall protector is 28 gauge sheet metal mounted on asbestos millboard, and it is even more effective if spaced out 1 inch from the wall to allow air circulation between the noncombustible surface and the wall. See Fig. 9-7. A radiant stove (one in which the firebox is uncovered by a jacket) may then be placed as close as 12 inches from the wall, and a circulating stove may be placed as close as 4 inches. Check local building and fire codes, however, because more clearance may be required.

Since the bottom of the stove is cooler than the sides, you do not need as much clearance between the bottom of the stove and the floor as you need between the sides of the stove and the wall. Almost all stoves are constructed on legs or some type of base, and whatever clearance they provide should be more than adequate. If your stove has no legs, set it on a noncombustible stand that will provide at least 4 inches of clearance between the bottom of the stove and the floor for air circulation.

Every stove must always be installed over a noncombustible floor material to protect the floor from glowing embers that may fall to the floor from the stove above. This floor covering should extend 8 inches to each side of the stove and 18 to 24 inches in front of the stove's door to protect the floor when fuel is added. A number of materials will protect the floor, including mortared bricks or stones, 24 gauge or thicker sheet metal, or asbestos millboard covered with sheet metal.

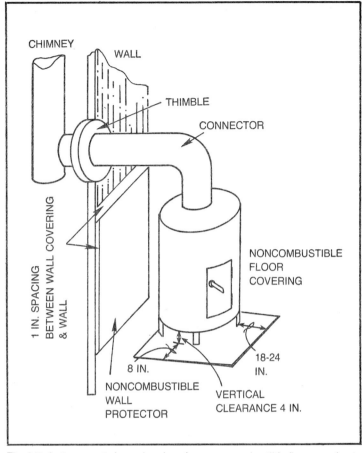

Fig. 9-7. A stove must always be placed on a noncombustible floor covering to protect the floor from heat and falling embers. Noncombustible materials must be placed on the wall if the stove is closer than 3 feet to the wall, and the wall protection is more effective if a 1-inch spacing is allowed between the wall protector and the wall. If the connector or chimney passes through a wall, ceiling or roof constructed of combustible materials, a thimble must be used to prevent a fire hazard.

Chimney and Connectors

No two combustible heating units should share the same flue, which means you will not be able to connect your stove to the existing flue from your operating fireplace or central heating system. If your house has an old fireplace chimney that is not being used, however, you may connect the stove to that. But before you use an existing chimney, have it inspected to be sure it is safe and will work for the type of fuel you plan to

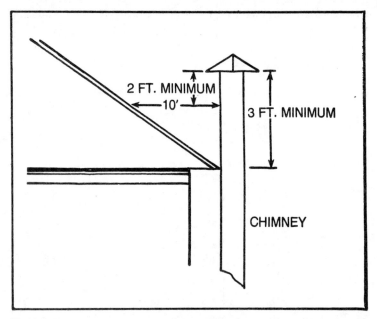

Fig. 9-8. For safety and proper performance the chimney must extend at least 3 feet above the roof at the point it intersects with the roof line, and must be at least 2 feet higher than any portion of the roof within 10 feet. The chimney also must extend at least 2 feet above any obstructions or walls within 10 feet.

use. If you plan to burn wood, you must be doubly careful of an old chimney because cresote will build up inside the chimney and will ignite on occasion. The chimney must be lined with a flue liner because the mortar in brick chimneys break down with time and a creosote fire can ignite the surrounding structure. Don't take chances!

If you do not have an existing chimney you can connect your stove to, it is quite easy to purchase factory-made chimney pipe from your stove dealer. These pre-cut lengths are constructed of sheet metal and asbestos, and are usually easy to fit together and install. You also might consider constructing a masonry chimney with a flue liner rather than purchasing the chimney pipes, but this is more costly.

The chimney must extend at least 3 feet above the roof at the point it passes the roof line, and it must be at least 2 feet higher than any ridge, wall or roof line closer than 10 feet. (Fig 9-8). The chimney does not necessarily have to extend above the peak of the roof, however. If the chimney passes directly above the stove through the ceiling and roof, you must

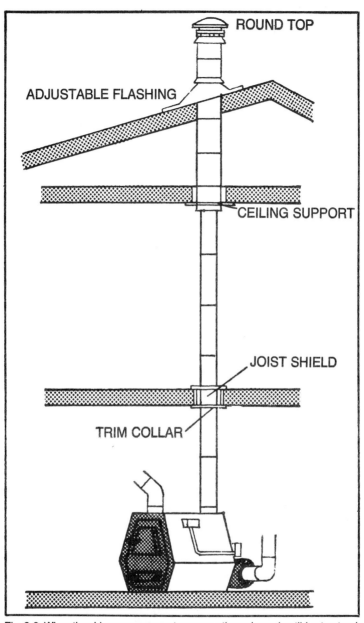

Fig. 9-9. When the chimney or connector are run through combustible structural portions of the house, spacers (or thimbles) must be used to provide clearance between the metal pipes and the surrounding wood. A wide variety of spacers are available for all types of stove installations. Where the chimney passes through the roof install an adjustable flashing to keep water from running down the hole cut for the chimney (courtesy Kickapoo Stove Works).

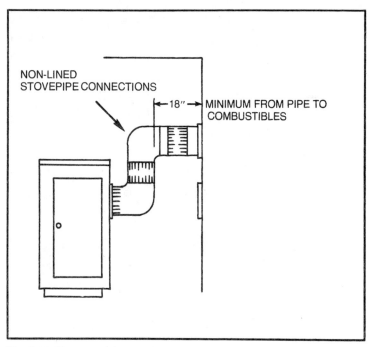

Fig. 9-10. Measurements of the spacing between the stove and nearest wall should be taken at the vertical stovepipe joint nearest the wall, unless the stovepipe is asbestos lined.

leave a 2-inch clearance between the chimney pipe and combustible structural materials. (Fig. 9-9). Flue collars and ceiling supports are available for these applications. Where the chimney intersects with the roof install an adjustable flashing to hold the chimney in place. This provides adequate clearance between the chimney and combustible roof materials and stops water from entering the hole around the chimney.

You may want to extend the chimney up the side of the house rather than through the ceiling and roof. Connect the chimney to the house structure with brackets that provide a 2-inch clearance between the pipe and the wood structural materials.

The chimney may be angled and offset if necessary to take it around obstructions on its path upwards, but angles of more than 30° should be avoided. Chimney pipe elbows are available to angle the chimney in any direction you want.

The connector, or stove pipe, attaches the stove to the chimney. This connector may be double-wall asbestos-lined pipe similar to chimney pipe, or it may be single-wall metal stovepipe. If single-walled stove pipe is used, you must be careful to measure the permissible distance between the stove and the wall as the distance between the closest side of the stovepipe and the wall, as in Fig. 9-10.

An asbestos-lined pipe passing through a wall or ceiling requires a 2-inch minimum clearance between the pipe and combustible materials. If a single-wall connecting pipe is used, you will have to allow a much larger spacing around the pipe. This can be achieved by cutting a hole in the wall or ceiling three times larger than the diameter of the stovepipe. A thimble will be installed in the hole, and the pipe will run through the center of the thimble. See Fig. 9-11. The thimble should have air holes in it to reduce the danger of fires from heat radiation. If you run a 6-inch single-wall stovepipe through a wall, figure on an 18-inch diameter thimble to run the pipe through.

Before you install your stove, check with your stove and fireplace supplier. He may have some good suggestions and installation tips, especially for the stove model you have purchased.

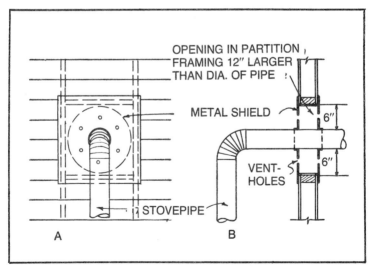

Fig. 9-11. When a single-wall stovepipe connector passes through a wall before joining the chimney, you must protect the surrounding combustible with a metal thimble as shown here. (A) front view. (B) side view (courtesy U.S. Dept. of Agriculture).

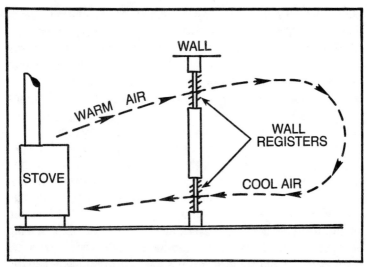

Fig. 9-12. Wall registers available at a furnace supply shop can increase the heat transfer between the room housing a stove and adjoining rooms. The warm air will rise and will pass through the high-wall register. Cool air will return to the stove through the low-wall register. A carefully placed small fan and an open door between the rooms will also help increase the heat flow.

Test Fire

Before you finally connect the flue and chimney in place permanently, you should build a test fire in the stove so you can check the chimney and connector for leaks and improper drafts. Build a small, smoky fire in the stove using paper or another suitable material. Now check the top of the chimney. The smoke should rise in the chimney if the drafting is proper.

Now close off the top of the chimney so that the smoke remains in the chimney. Examine each chimney and connector joint for smoke leaking out. Any leaks will have to be sealed before the stove can be used.

Reopen the top of the chimney and allow the smoke to escape. If the smoke does not rise through the chimney but remains in the stove, there is an improper downdraft that must be corrected.

OTHER STOVE INSTALLATION IDEAS

Since your stove is not a part of the central heating system of your home, one of your major problems in utilizing wood heat is to get the heat from the stove into adjoining rooms. Without placing the stove's heat directly into a duct system, this can be somewhat difficult, although with a bit of ingenuity you can come up with several ways to distribute the heat about your house. If your stove is a small one-or two-room stove, this may not be a major concern. But if you have a five to seven room stove, you want to distribute that heat!

Purchasing a stove with a built-in blower unit will help a lot. These blowers remove the heat from the stove and push it out into the living area where where it can do some good. This blower will probably be inadequate to distribute the heat through the entire house, so you should place some small fans about the house to direct the warm air into the rooms desired. Since warm air rises, the most effective location for such fans is near ceiling level—especially in the room in which the heater is located.

You should attempt to place the stove in centrally located rooms—the rooms that are used most in the winter months. Attempt to "expand" the room the stove is placed in by opening doors to adjoining rooms. You might consider placing wall registers in the walls between adjoining rooms in order to increase air circulation and heat flow into other rooms (Fig. 9-12).

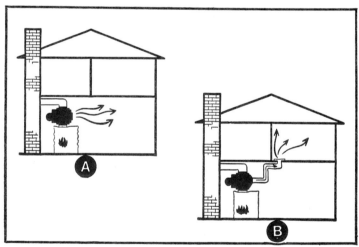

Fig. 9-13. The flue heat exchanger is a simple add-on item to the flue of your combustible fuel stove or furnace that captures some of the heat escaping up the flue. This heat may be directed out into the room housing the stove as in (A), or it may be ducted to a separate room as in (B). Adding the flue heat exchanger to your stove or furnace will not change the amount of heat the stove produces by radiation or normal convection currents. The stove's normal operation remains unchanged (courtesy Dolin Metal Products).

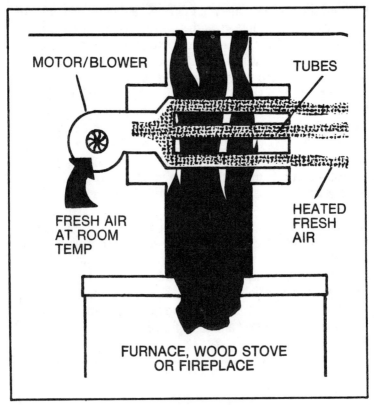

Fig. 9-14. Flue heat exchangers work like all other heat exchangers. Hot air and gases enter the heat exchanger's tubes and warm the walls of the tubes. Fresh air blown over those tube walls removes heat and takes it to the living area (courtesy Dolin Metal Products).

A flue heat exchanger attaches to the stove flue as it leaves the stove and captures some of the heat escaping up the flue. This additional heat can be sent out into the room with the wood stove, or it may be ducted into another room (Fig. 9-13).

There a number of these units available today, but their operating principles are similar. The unit simply attaches to the flue as an "add on" item. The gases exiting the flue are now sent through the heat exchanger, and the gases warm the heat exchanger walls. The unit's fan blows air through the heat exchanger, and the air picks up heat from the walls. The air does not come in contact with the flue gases. See Fig. 9-14. The addition of a flue heat exchanger has no effect on the basic

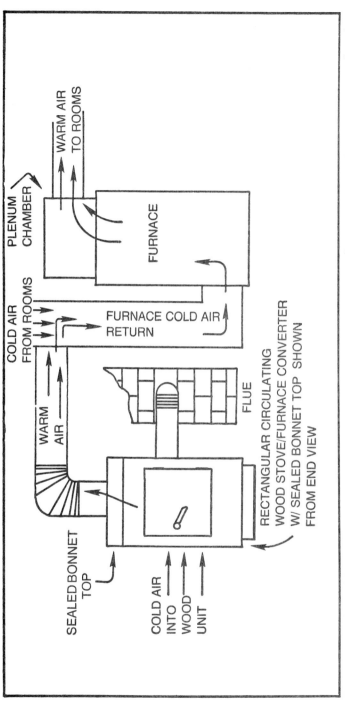

Fig. 9-15. A wood or coal stove can easily be connected to your existing central heating system if a jacket and closed bonnet are used to direct the heat flow into the central duct system. A system such as this is essentially a wood furnace. The hot air from the wood stove is pushed into the cold air return duct of the central heating system. The heat is then distributed throughout the house through the duct system.

operation of the stove, which continues to produce and distribute heat in the same manner as before. The flue heat exchanger simply recovers some of the heat that otherwise would be lost up the flue.

Most of these units are easy to add to any stove or furnace and come equipped with their own electric motor.

ATTACHING A STOVE TO YOUR CENTRAL HEATING SYSTEM

If you already have a central heating system and want to utilize its distribution capabilities to get the most out of your stove you can connect the stove into your central heating system, as in Fig. 9-15. Using this installation the stove's heat is sent into the furnace duct system, where it is sent evenly to each and every room in your house.

A diagram of a wood-burning stove designed for this purpose is shown in Fig. 9-16. If you compare this diagram to a similar stove model shown in Fig. 9-1, you can see that the major difference between the two units is that the "furnace converter" model has a sealed bonnet on top. This bonnet keeps the heated air from flowing out of the stove and directly into the room. Instead, the heated air is directed through the vent at the top of the bonnet, where a duct takes the heated air to the central furnace's cold air duct (also known as return air duct).

If you have space in your furnace room for one of these units, their installation is fairly simple. A duct is run from the stove unit to the cold air return of the furnace, following the manufacturer's instructions. Installing ductwork is discussed in Chapter 8. If you do not have space in your furnace room for such a unit you can place the stove in a separate room and duct the heated air to the furnace.

All other installation procedures for these units are the same as for a stove. You must be sure to allow proper clearance between the stove and wall, and you must provide proper wall protection. **Important: the stove can not share the same flue with the central heating furnace.** Doing so creates a safety hazard. You must vent the stove flue to a separate chimney.

The units shown in Figs 9-15 and 9-16 are essentially circulating stoves that have been modified slightly to work as a

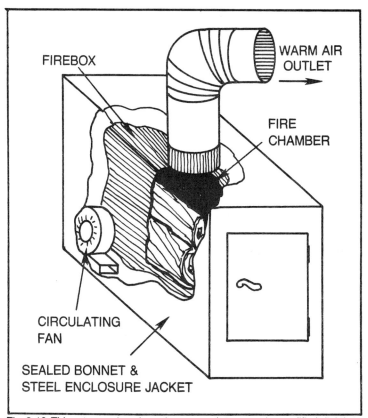

Fig. 9-16. This cutaway view shows how a wood stove can be modified slightly to connect to a central heating duct system. The chief difference between this stove and the one shown earlier in Fig. 9-2 is the sealed bonnet on top of the unit. This captures the heat, which is ducted to the central heating system.

furnace and to connect to the central heating system. They are not, strictly speaking, wood or coal furnaces because they do not have a large fan to distribute heated air through a duct system and they are designed primarily as stoves. You can install a similar system by connecting a small wood furnace to your central heating system.

Figures 6-4, 6-5 and 9-17 show other installations of wood circulating stoves or small wood furnaces in line with existing furnaces.

STOVE MAINTENANCE

Compared to many heating systems, the stove is relatively maintenance-free. Ashes should be removed about once a day—*always* in a metal container. Never store ashes in a combustible box! Before you retire your stove for a season you should remove all the ashes, since the presence of ashes may accelerate the rusting process of your grate.

Your stove's grate will rust with normal use, and eventually it may rust through. If it does, replacement grates are easy to obtain. If any part of the firebox should ever rust through, have it repaired immediately. A metal shop or furnace shop will usually be able to do the repair work for you.

Once a year, check the connectors, chimney pipes and points where the pipes pass through combustible walls or ceilings. This is especially important when the connector is made of single-wall stovepipe that can rust quickly and create a fire hazard. Replace any sections that have rusted through.

If you burn wood, creosote will build up on the sides of the chimney flue. The best way to handle creosote is to prevent the problem in the first place—burn seasoned hardwood and avoid unseasoned wood or softwood. If creosote builds up enough to restrict the flue, you may be able to remove part of the buildup by using a chemical cleaner sold for this purpose. To remove creosote high in the chimney, suspend a weighted object on a rope down the chimney and scrape the walls of the chimney with the object.

WOOD AND COAL FURNACES

Once a wood stove is enclosed in a jacket and it directs the heat through a duct system into the living area, it is essentially a wood furnace. Thus, the previous section showing how to connnect what is basically a wood stove to your central heating system actually describes one type of wood furnace. There are many different types of wood furnaces available. The simplest types are like the one previously mentioned: a wood stove enclosed in a jacket with the air directed through a duct into an existing central heating system. Different manufacturers call such systems by different names. Some call them "stoves," some call them "furnaces" and others call them "furnace converters."

The more complex wood and coal furnaces have a number of features that make them more like what we usually think of as "furnaces." They are designed to be the only heating source in a central heating system and to be connected

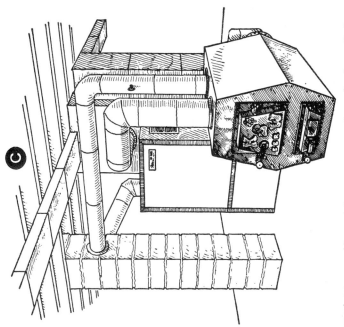

Fig. 9-17. This small wood or coal-burning furnace is large enough to heat several rooms, and it may take over as your primary heating unit when connected to the existing central heating duct system. This unit can be connected to the duct system alongside your present furnace. It is connected by running a duct from connector (A) to your furnace duct system. (B) is the air filter (courtesy Kickapoo Stove Works.), (C) shows the furnace connected to the duct system.

directly to the duct system. They have fans sufficient to distribute the air through the entire duct system. They also have electric thermostats that adjust the fuel burning rate according to the temperature in the living area. They are designed to be operated in a remote location, removed from the living space.

Almost all of these units will burn either wood or coal. Most will hold about a 12-hour supply of fuel in their fireboxes. Some coal-burning units will connect to a stoker that will automatically feed the firebox, reducing even further the amount of attention the furnace requires.

Wood and coal furnaces have basically the same components that will be found in any combustible fuel furnace, such as a gas furnace. They are usually somewhat larger because of the large firebox necessary to store and burn the fuel. But the components are the same: combustion chamber (firebox), heat exchanger, plenum chamber, duct system and fan.

Once you begin considering installing a wood or coal furnace in your home, you will discover that choosing among the wide variety of styles, sizes and designs is not easy. First you must decide just what purpose you want your furnace to serve. If you want it only to be an add-on furnace unit to your existing central heating system, a relatively small capacity furnace/stove with a small blower fan will probably be just the ticket. For most homeowners, such a system will probably be sufficient to take advantage of lower wood heating costs, yet retain the flexibility of having a conventional-fuel furnace available when it is needed. Such systems are also less expensive than adding a larger furnace.

But you may be thinking of replacing your present gas, oil or electric furnace. If so, you can purchase a wood or coal furnace of the same capacity as the furnace you are now using. If you heat with hot water or steam, wood or coal-burning boilers are also available.

One type of furnace you certainly should consider is a "combination" furnace, like the one shown in Fig. 9-18. These furnaces are made by a number of wood and coal furnace manufacturers, and they allow you to use gas or oil as a secondary fuel. These furnaces have greater flexibility than a wood-only furnace, but they are also more costly. Some of these furnaces burn a small amount of gas or oil even when wood is burning. Others switch only to gas or oil fuel if the wood fuel is insufficient to heat the home.

When purchasing a wood furnace, you must consider how easy the unit is to operate. You also may want to consider whether it will operate during power outages. Whether it requires a secondary fuel to burn the wood or coal may be a consideration. The larger the firebox and opening, the larger the logs the furnace will accept. This means less work for you in sawing the wood.

Fig. 9-18. This combination furnace, or multifuel furnace, burns oil as well as wood or coal. This furnace model uses the secondary fuel (oil) continually while burning the wood. This burning process turns the wood to charcoal and eliminates air from the wood combustion process. (A) the flue heat exchanger that recaptures some of the heat escaping in gases up the flue. (B) the oil burner unit. (C) ash pit door. (D) refueling door. Courtesy Longwood Furnace Corp.

You must carefully consider where you will place the furnace, since it probably will be heavy enough to require special floor supports or foundation support. Talk with your furnace dealer to get help in planning modifications to your house before you buy a wood furnace.

FIREPLACES

Fireplaces, like wood stoves and furnaces, have enjoyed a sudden increase in interest in recent years as homeowners have become more energy conscious. At one time, fireplaces were an important source of heat for most homes in the nation. But the use of modern furnaces burning gas and oil reduced the fireplace's usefulness as a heat source, and most fireplaces were installed for their visual appeal. Other fuels were cheap then, homeowners reasoned that there was no need to go through the bother of tending a fire in the fireplace when they could let their furnace do the work.

Today, of course, other fuels are no longer cheap, and homeowners searching to cut their fuel bills have turned to the fireplace for help. There are a couple of problems with depending on the fireplace as a supplementary heating source, however. The fireplace is a very inefficient heating unit by its very nature. In some fireplaces, as much as 90 percent of the heat produced by the burning goes up the flue.

Fireplaces consume large amounts of air in burning wood, and this air usually comes from the living area of the house. The air extracted for the burning process must be replaced by cold air from outdoors, and this new air must be heated by the furnace system. When you consider this "air extraction" process with the relatively small amount of heat the fireplace is producing, it is easy to see how a fireplace can actually raise, instead of lower, your furnace's heating bill.

A large number of advances have been made recently in fireplace design and installation to dramatically improve the fireplace's performance as a supplementary heating unit. But you must remember that many fireplaces built into homes more than 10 years ago were built primarily for the visual appeal of a fireplace, and scarcely a thought was given to their heating performance. Such fireplaces incorporate none of the efficiency-increasing designs, and operating them will likely increase your heating bills considerably.

Later in this chapter we will discuss some of the ways you can increase the efficiency of your fireplace but one point should be made here. No matter how many of these devices you incorporate into your fireplace, you will probably not be able to extract the same efficiency from your fireplace as you can from a modern-design wood stove. And usually, the wood stove will be much cheaper! If you prefer the fireplace for its visual appeal, that's a perfectly rational preference. But you should also recognize that the visual appeal of a fireplace is costing you a lot of wasted heat sent up the chimney.

Efficiency figures are extremely uncertain, but as a rule of thumb, you can figure that a fireplace with no efficiency-increasing devices is about 10 percent efficient. Adding hollow-tube grates or installing a heat-circulating fireplace (discussed later in the chapter), may increase the efficiency to as much as 25 to 30 percent. A manually controlled wood stove, on the other hand, costs much less than a fireplace and is about 20 percent efficient. A modern circulating stove with a thermostatic control and a blower unit is also considerably less expensive than most fireplaces, and has an efficiency of about 40 to 50 percent.

As you can see, if inexpensive heat for your home is your major consideration, a stove offers some large advantages over a fireplace. However, if you like the appearance of a fireplace, there are a number of ways you can use that fireplace to help heat your home.

PARTS OF A FIREPLACE

Figure 9-19 is a diagram of a fireplace showing the basic parts of the fireplace. The heart of the fireplace is the fire chamber, where the wood burns. In traditional fireplace designs the fire chamber is lined with fire bricks for a masonry construction. But much better heating is obtained when a double-walled steel liner of a heat circulating fireplace system is used as the fire chamber. Heat circulating fireplaces are discussed later in the chapter.

The walls of the fire chamber are sloped inward toward the back of the fireplace to increase the amount of heat that radiates into the living area. Very little heat would reach the living area if the sides went straight back.

As you can see from the diagram, the top of the fire chamber slopes inward and upward toward the living area to direct radiant heat. At the top of this slope is the fireplace's throat, which can be closed with the damper. The damper is closed when there is no fire burning in the fireplace to prevent drafts through the house and up the chimney. If the damper is not closed, warm air inside the house will escape up the flue.

Improper use of the damper is one of the biggest sources of fireplace inefficiency and lost heat. If the damper is left open when there is no fire in the fire chamber, you will waste a lot of heat and money. One of the big advantages of installing glass doors on a fireplace opening (discussed later in the chapter) is these doors can close off the flue from the living area while a fire is burning low. The damper does not have to be tended nearly as closely when glass doors are used because the closed doors will prevent most of the problems cause by an open damper.

The smoke shelf deflects cold downdrafts, which can be prevalent during the early stages of a fire. The smoke chamber and smoke shelf combine to prevent downdrafts from carrying smoke back into the living area.

The flue is lined with a masonry flue lining that extends to the top of the chimney. If you have two fireplaces or a fireplace and a stove, the flues may run through the same chimney, but each combustible fuel unit must have its own flue. The flue must be sized properly for the size fireplace opening used. A fireplace supply dealer will have charts to show you how large the flue should be for any size fireplace opening.

The flue runs through the chimney to the top of the chimney. It must extend at least 2 feet above any part of the roof or house walls closer than 10 feet, as discussed earlier in the section of this chapter on installing wood stoves.

The hearth is the floor of the fire chamber. It is built of fire brick or other heat resistant materials, and it extends in front of the fireplace opening to protect the floor from heat. The hearth should extend about one foot to the side of the fireplace opening and about 1½ feet in front of it.

Some fireplaces have a metal door in the hearth that opens to let the homeowner push the ashes into an ash pit under the fire chamber. The ash pit occupies the space between the fireplace foundation walls. Another small metal door at the wall of the fireplace foundation allows the ash pit to be cleaned.

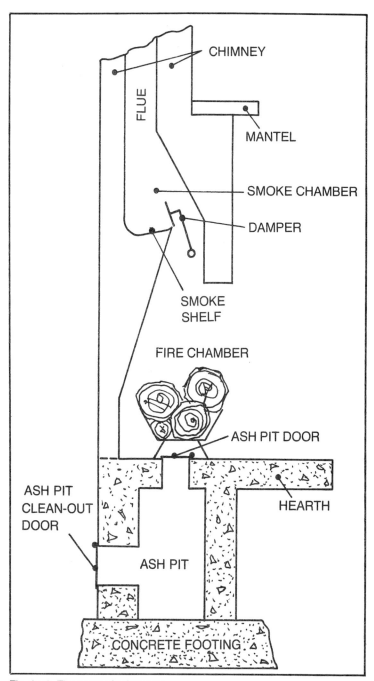

Fig. 9-19. The parts of a fireplace.

One of the most common complaints from fireplace users is that their fireplaces are too small. The larger the fireplace is, the larger the fire you'll be able to build in it. Also, larger fireplaces will accept much larger logs for burning. If your fireplace can burn large logs you'll have less work cutting and splitting wood into smaller sizes.

If wood in your area is inexpensive, or if you have located a source of free wood, you'll probably be burning rather large fires to save more money on your heating bill. If your main reason for installing a fireplace is to save on your heating bill, you will probably want to install a 36-inch (measured across the front opening) or larger fireplace.

FIREPLACE SAFETY

Before finalizing construction plans for a fireplace in your home, always check with local building code authorities and local fire department officials to get a copy of fireplace restrictions and installation rules. Local codes vary, and they can change frequently. Don't trust your memory or your neighbor's word on what the codes say. Do it right. Get a copy of the codes before you begin to avoid having to do the job over again when it won't pass inspection!

There are a number of safety precautions you must take in fireplace installation. Usually these will be prescribed by a building or fire code in your locality. If they are not, however, the following recommendations serve as a general guide.

There should be no combustible materials within 4 inches of the fireplace opening. If you plan to install a wood mantel that will extend more than 2 inches out from the wall, it must be at least 12 inches above the fireplace above the fireplace opening.

The chimney must have a flue lining to prevent the chimney from becoming too hot and igniting nearby combustible framing boards. There should be a 2 inch air space between the fireplace or chimney and any combustible structural materials.

Metal parts such as the damper or combination damper-smoke chamber should be well constructed. The damper should be reasonably easy to open and close, but it must not move so easily that it might move by itself. The damper must seal tightly.

The traditional type of fireplace is the masonry fireplace. This fireplace is constructed almost entirely of bricks, mortar, concrete and other masonry products, and it can be installed in a home only after a foundation has been set for a fireplace.

PREFABRICATED FIREPLACES

The prefabricated fireplace eliminates the need for most of the masonry and foundation work required for a masonry fireplace. The prefabricated fireplace is built of metal instead of bricks and mortar, and it may be purchased as an entire kit that includes all the parts of a fireplace: metal flue chamber, flue and chimney, damper, hearth and even fireplace trim. These fireplaces are installed much like a wood or coal stove. You bring the fireplace unit into your home, place it where you want, and hook it up!

Prefabricated fireplaces come in many designs, ranging from contemporary free-standing fireplaces to masonry-looking fireplaces that can actually be enclosed in a wood structure and trimmed with imitation bricks to look like a masonry fireplace!

There are a number of advantages to prefabricated fireplaces. One of the biggest advantages is almost any homeowner can install them himself. This saves costly labor expenses for masonry work. Most of these fireplaces can be purchased at any home decorating center with all necessary parts included, and the directions should be complete and easy to follow. Be sure to check this out before you buy if you plan to install one of these fireplaces yourself. If the directions are cryptic notations that can be understood only by persons completely familiar with installing the fireplace, they won't do you much good. You'll have a lot of problems putting your fireplace in.

Prefabricated fireplaces require no foundation and no footing. They can be installed right on the subfloor and against a wood wall. They are usually well-insulated or are built with double-wall construction to protect against burning the room structure. You should be sure to check these features before you buy.

Because they require no foundation and generally require no special support, prefabricated fireplaces can be installed virtually anywhere in a home—along an inside wall, in an upstairs room, even in the center of a room. These fireplaces usually come with their own flues and chimneys, but they may be connected to masonry chimneys by using connector pipes.

When purchasing a prefabricated fireplace, check it for solid construction, strong welds and convenience in use. If the fireplace sides will be exposed to the living area, as in a free-standing fireplace, check the sides for insulated double-wall construction that will not cause burns.

INCREASING FIREPLACE HEAT OUTPUT

The basic fireplace designs are pretty much the same today as they have been for decades, but there are a number of additions to most modern fireplaces that increase their energy efficiency tremendously. In fact, without the addition of one or more of these heat-producing devices, most fireplaces will increase heating bills rather then reduce them!

The traditional fireplace design is a masonry fireplace, a design that is still being installed in many new homes today. The all-masonry fireplace has a fire chamber constructed of firebricks. Virtually the entire heat output of the masonry fireplace comes from heat radiated into the living area.

Heat Circulating Fireplaces

The heat circulating fireplace, also called the circulator fireplace, is a large improvement over the standard fireplace because it combines convection air currents with radiation to transfer heat from the fire into the living area. The fire chamber of a circulator fireplace is a factory-formed double-wall steel unit that allows air to circulate behind the fire chamber, pick up heat from the fire, and exit into the living area (Fig. 9-20). Air circulation through the unit and heat output is increased with the addition of one or two small fans to draw room air into the air space behind the fire chamber.

The advantage of the circulator fireplace over a traditional masonry fireplace are apparent. Whereas a conventional masonry fireplace provides heat only from the radiation of the fire, the circulator fireplace provides the same radiated heat plus the additional heat from air circulation around the back of the fire. The circulator fireplace greatly increases the fireplace's heat output by capturing much of the heat that otherwise would escape up the flue.

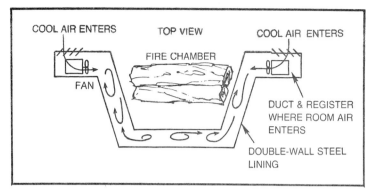

Fig. 9-20. The circulator fireplace draws air from the living area at the bottom of the fireplace and circulates it through a double-wall fire chamber lining. The air picks up heat from the inside wall of the lining, and the warm air exits at the top of the fireplace lining to return to the living area. This recaptures a large amount of heat that otherwise would escape up the chimney. Electric fans are often included in circulator fireplaces to increase the warm air flow.

An additional advantage of the circulator fireplace is that it is much easier to install than a conventional masonry fireplace. A masonry fire chamber requires careful measuring, shaping and fitting for the brick lining. But the circulator fireplace is a steel form that combines a shaped fire chamber, throat, smoke shelf and damper, all ready to install. The bricks are then added around this form, as in Fig. 9-21. An able do-it-yourselfer with some masonry skills can install a circulator fireplace himself.

Glass Doors

Glass doors are another energy-saving feature of modern fireplaces. Wire screens have long been placed in front of burning fires to stop sparks from popping out of the fireplaces and into the living area. Glass doors provide this safety, plus another important feature. They block the fireplace opening so warm air inside the house cannot be sucked up the chimney. Because an open fireplace draws warm inside air through the fireplace opening and sends it outside through the chimney, an open fireplace can actually ADD to the furnace heat bill instead of lowering it. Glass doors prevent some of this warm air loss by controlling the amount of air that is drawn from the living area into the fire chamber.

Glass doors also allow you to leave the fireplace damper open at night while the fire burns down. You can just close the

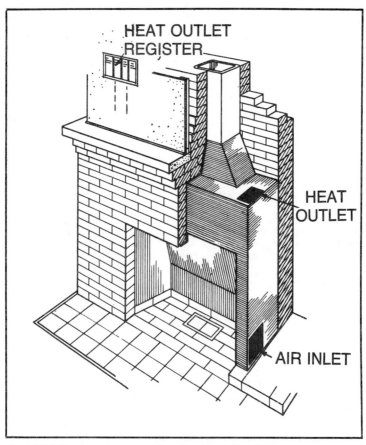

Fig. 9-21. This shows how a circulator fireplace is installed. The steel fireplace form is put in place and brickwork is installed around the form. The air is drawn into the unit at the air inlet at the bottom. Air circulates around the back and sides of the fire before exiting at the heat outlet at the top. With the use of ducts and registers, this heat outlet may be placed almost anywhere in the home the homeowner wants (courtesy U.S. Department of Agriculture).

doors and go to bed while the fire continues burning. When the fire goes out, the closed doors will prevent much warm house air from being drawn up the flue. Glass doors can easily be used with other efficiency-increasing devices, and are frequently used with circulator fireplaces.

Hollow-Tube Grates

Hollow-tube grates (Fig. 9-22) perform much the same function as a circulator fireplace in that air is drawn in from the living area, circulated around the fire and warmed, and then returned to the room. You can purchase models of these grates that use either natural air currents for circulation or electric circulating fans.

Hollow-tube grates may be used with a circulator fireplace to increase the heat output more than either unit operating alone would be able to increase it. If you plan to use hollow-tube grates with glass doors, however, you must purchase a glass door assembly with the grate built in as part of the assembly. Otherwise, you will render the tubes ineffective when you close the glass doors.

When purchasing these grates, be sure you get a heavy-gauge metal grate assembly. The tubes should be constructed of at least 14 gauge or 12 gauge material (lower numbers mean thicker-walled metal tubes). Lightweight grates will quickly burn out.

Fig. 9-22. Diagram of a hollow-tube fireplace grate (courtesy Shenandoah Manufacturing Company).

Outside Air Supply

Burning wood in a fireplace requires a lot of air, and one of the easiest places for the fireplace to pull this air from is the living area. This presents a dilemma for the homeowner trying to conserve energy. The fireplace is drawing air from the living area, and this air is being replaced by colder outside air that enters the house by infiltration. This infiltration air must be heated, and that adds to the heat load of the house.

If the house is sealed tightly so that little infiltration air can enter the structure, there is a risk of depleting the oxygen supply and endangering the house's inhabitants. Or the fire may begin drawing air down the chimney to supply it with air.

There are a number of possible solutions to this problem. One of the simplest is to open the ashpit doors to allow the fire to draw air needed for combustion. When this is done, the fire is drawing outside air into the fire chamber instead of warm inside air. You must remember to close the ashpit doors when there is no fire in the fireplace. Otherwise, the cold outside air will enter the house.

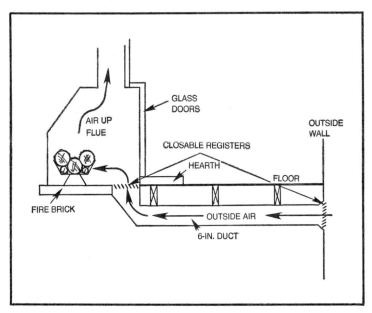

Fig. 9-23. An outside air supply to the fireplace saves energy by allowing the fire to consume cold outdoor air in burning instead of warm, inside air. This duct must have registers which close to prevent the outside air from entering the fire chamber when the fire is not burning.

If your fireplace does not have an ashpit, you can run an outside air duct under the house to the hearth in front of the fireplace, as in Fig. 9-23. This achieves the same result of allowing the fireplace to draw its draft from unheated air instead of sending warm inside air up the chimney and replacing it with cold air. The outside air duct should be installed with an adjustable register that can be regulated as needed and closed when there is no fire.

CHECKLIST

1. Will wood or coal provide you more heat per dollar into your home than other fuels?
2. For least amount of heat sent up the flue and best efficiency, your wood and coal-burning unit should have a thermostatic damper control and a convection system of heat transfer (circulating design).
3. Flue heat exchangers recapture heat otherwise wasted up the flue.
4. Burn dry, seasoned hardwood.
5. Distribute heat produced by stove and fireplace throughout your house through fans or duct system.
6. Do not allow creosote to build up in flues.
7. For fireplaces, keep damper closed when fire is not going in fire chamber. Keep glass doors closed.
8. Use high efficiency fireplace designs, such as circulator fireplaces, hollow-tube grates, glass doors and an outside air supply.

Heat Pumps

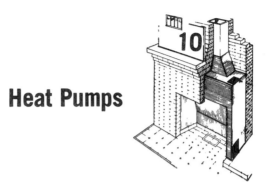

Heat pumps have been around for home heating and cooling installations for more than two decades, but few homeowners had even heard of them until a few years ago. Now, it seems, almost every homeowner has heard of the heat pump. He's heard satisfied users and heat pump dealers swear by it; he's heard complaints from other users and advice that only a fool would install such a contraption. No wonder homeowners and home builders are confused about the heat pump! It's a device that's much talked about, but seldom understood.

In this chapter, we show how a heat pump operates, and how it may be able to reduce your heating costs. We also will give you some tips on purchasing a heat pump and give you advice on when not to purchase one. Finally, for heat pump owners, there are sections on maintenance and servicing.

INCREASING POPULARITY

If you've been watching television or reading newspapers and magazines recently, you've undoubtedly run across a number of news stories and advertisements for heat pumps. Heat pumps have received a good deal of attention lately, and much of it began in 1973 when heating oil and gas prices skyrocketed. Suddenly, the heat pump's operating cost became competitive with oil and natural gas in many regions, and many homeowners began installing heat pumps in new homes as an energy-conservation measure that would reduce their heating costs.

The big reason for all this sudden interest is the heat pump's operating efficiency—often more than twice that of an electric furnace. This makes them an especially attractive alternative to electric resistance heating in those parts of the United States where natural gas supplies are dwindling and no new natural gas customers are being accepted. And in many localities, heat pumps now cost less than oil or gas furnaces to operate.

Although heat pumps can save many homeowners money, they would not have become so popular if manufacturers had not made a number of needed design improvements in heat pumps. It's only been within the last several years that improved heat pump design has made the heat pump a viable alternative in the northern half of the country. Early heat pumps were often unreliable. They were very inefficient at temperatures below freezing, and special components designed to withstand the rigors of heat pump operation had not been perfected.

But most of these problems have been corrected, and the heat pump is coming on strong. New components designed especially for heat pumps have taken care of the reliability problems of a decade ago, and many heat pumps are now designed to operate efficiently in many northern climates. The heat pump is an important part of the current energy scene, and experts say it is destined to play an even larger role as fossil fuel prices rise and supplies dwindle.

WILL A HEAT PUMP SAVE YOU MONEY?

While it may be true that thousands of homeowners have saved money on their heating bills by installing heat pumps, heat pumps are not for everybody. Often, other heating fuels are so inexpensive that the cost of operating a heat pump over an entire winter is about the same as a gas or oil furnace. Also, the cost of replacing an existing heating system with a heat pump is often prohibitory, so that if you already have a heating system installed in your home you may not be able to convert economically to a heat pump. But saving money by installing a heat pump depends on a number of factors that are different for each homeowner, and you must consider your individual situation to determine whether it would be wise for you to purchase a heat pump.

You first must compare the heat per dollar you would get from a heat pump against the heat per dollar from other types of fuel available to you. Chapter 3 outlines how to do this. In doing this, you will have to account for the fact that heat pumps vary widely in efficiency according to the winter climate. The milder the climate, the more efficiently the heat pump operates. Thus, if you live in a cold winter climate, the heat pump may not operate efficiently enough to save you money over gas or oil furnaces. The effect of climate on the heat pump efficiency is discussed later in this chapter.

Because the heat pump is both a heating unit and an air conditioning unit, you should consider installing a heat pump only if you are going to install central air conditioning. If you prefer room air conditioning or no air conditioning, you would be better off to simply install a heating-only furnace that would cost much less than a heat pump. The additional cost of a heat pump above a heating-only furnace would be so great that it would be almost impossible to recoup the extra cost through reduced heating bills.

By the same token, if you are considering installing a heat pump to replace your present worn-out furnace, the heat pump is probably not an economical alternative unless your air conditioner needs replacing, too, or unless you can find a buyer for your used air conditioner.

The heat pump is a central climate control unit, which means you must have insulated ductwork to carry the warm and cool air to the rooms. If you are considering installing a heat pump in an existing house that has no ductwork, you must consider the additional cost of installing the necessary ducts.

Heat pumps generally cost more to purchase than conventional heating and cooling systems, and you must take this additional initial investment into account when figuring the heat pump savings. You can expect to pay about 10 to 25 percent more for a heat pump installation than you would pay for a conventional central heating/air conditioning system.

On the other hand, heat pumps can save money for many people, since the heat pump's efficiency is reflected directly in their utility bills, Heat pumps are an especially good choice for those who are planning to install electric resistance heating furnaces with central air conditioners. On the average, heat pumps save about 20 percent in operating costs over such systems, and in some parts of the country the savings can reach 30 to 40 percent. But if natural gas is available, it is probably less expensive to operate a natural gas furnace than a heat pump (at least at current energy prices).

HEAT PUMP OPERATION

Conventional furnace systems use electricity, gas, oil, wood or coal to generate heat, but the heat pump uses principles of refrigeration to transfer heat from one location to another. The heat pump does not use electricity to generate heat in its basic operation, as does an electric furnace. Instead, it uses electricity to operate a compressor that pumps refrigerant through the system. This refrigerant absorbs heat from the outside air and transfers it to the air indoors.

It may seem odd that on a cold winter day a heat pump could actually remove heat from the outside air, but that's what a heat pump does. Actually, even the coldest winter air contains some available heat, and as long as the temperature of the outside air is above the temperature of the evaporator coil outdoors, heat can be removed.

The heat pump works as an air conditioner during the summer months, removing heat from inside the living area and depositing it outdoors. During the winter when the heat pump is on its heating cycle, it reverses so that it is essentially operating as an air conditioner in reverse. On the heating cycle it removes heat from the air outdoors and deposits it inside the house.

You might visualize this by thinking of a window air conditioner that is cooling during the summer. Air blowing over the indoor coil, the evaporator, is cool, but air blowing across the outdoor coil, the condenser, is warm. Heat inside the house is being transferred outdoors and dissipated. If you were to turn the air conditioner around in the window, you would have the warm air blowing inside as the evaporator removed heat from the outside air. This is how a heat pump works on the heating cycle.

The heat pump operates on basic refrigeration principles. Its main components are an evaporating coil, a condensing coil, a refrigerant that fills the system and changes from a

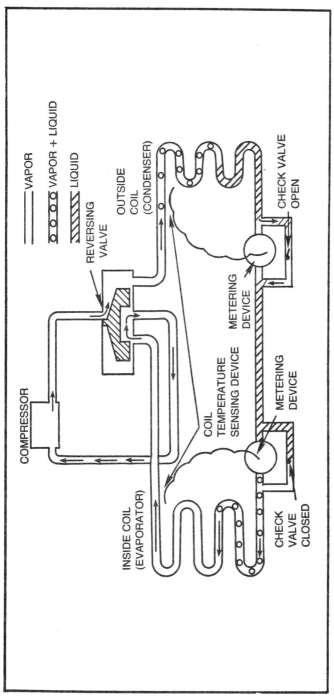

Fig. 10-1. The cooling cycle of a heat pump. The indoor coil, the evaporator, absorbs heat as the refrigerant boils and changes from a liquid to a gas. The outdoor coil, the condenser, releases heat into the outside air as the refrigerant is compressed in the coil and changes from a gas to a liquid. The metering device and check valve nearer to the evaporator restrict the flow of refrigerant through the lines, making pressure in the condenser possible. The compressor pumps the refrigerant through the system and places the refrigerant under pressure in the condenser.

liquid and a gaseous state to absorb and release heat, a compressor that pumps the refrigerant through the system and helps the refrigerant absorb heat, a reversing valve that switches the direction the refrigerant flows through the system to change the roles of the two coils, and fans that move the air over the coils and through the duct system to aid in heat transfer.

Figure 10-1 shows the summer air conditioning cycle of the heat pump. The indoor coil, which is the evaporator during the cooling cycle, picks up heat from air passing over the coil. This happens because liquid refrigerant entering the evaporator coil "boils" to form refrigerant vapor. This boiling process absorbs a lot of heat from the surrounding air, and makes the evaporator coil feel cool. Heat pump refrigerant boils at sub-zero temperatures.

The refrigerant, which is now a vapor, travels to the compressor, and the compressor pumps the vapor into the condenser (the outside coil on the cooling cycle). At the condenser the refrigerant is placed under pressure. Under pressure from the compressor, the refrigerant vapor returns to a liquid state as air blowing over the condensing coil removes heat from the refrigerant. Thus, the refrigerant "condenses"—changes from a vapor to a liquid—inside the condenser, and this process gives off a good deal of heat. This heat, which was absorbed by the refrigerant in the evaporator, is blown into the outside air by the condenser fan.

Figure 10-2 shows the heat pump heating cycle. The reversing valve changes positions to switch the direction refrigerant flows through the coils, and this changes the functions of the indoor and outdoor coils. Notice, however, that the direction of refrigerant flow through the compressor is unchanged.

Now the indoor coil is the condenser, dispensing heat from the refrigerant into the living area. On the cooling cycle, the indoor coil was the evaporator. Now the outdoor coil is the evaporator, absorbing heat from the outside air. On the cooling cycle, the outdoor coil had been the condenser.

Now the refrigerant boils in the outdoor coil to absorb heat from the outdoor air. Because of the low boiling temperature of refrigerant, the refrigerant will absorb some heat on even the coldest days. The compressor carries the refrigerant

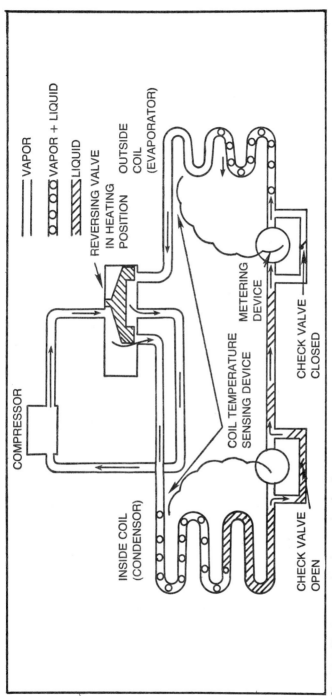

Fig. 10-2. The heat pump heating cycle. The reversing valve has changed positions to change the direction the refrigerant flows between the two coils. The direction the refrigerant flows through the compressor has not changed, however. The outside coil is now the evaporator, and the refrigerant boils there to absorb heat. The inside coil has become the condenser, where the refrigerant is compressed and changes into a liquid to release heat.

vapor to the inside coil, now the condenser, where the refrigerant condenses from a gas into a liquid as the fan removes heat from the coil and sends it into the duct system.

The heat pump units and their parts are shown in Figs. 10-3 and 10-4. This type of system—where each coil and its components is housed in a separate unit, one indoors and one outdoors, is called a "split system."

Check Valves and Metering Devices

Restricting the flow of refrigerant through the system is essential to any refrigeration system, and the heat pump is no exception. The restriction must be in the refrigerant lines just before the refrigerant flows into the evaporator. Because of the restriction in the refrigeration lines at this point, the refrigerant flowing toward the evaporator will be under pressure, and this pressure keeps it in a liquid state. A small amount of refrigerant is allowed to flow through the restriction into the low-pressure evaporator. Inside the evaporator this refrigerant, no longer compressed into a liquid, will expand and boil to form a vapor once again.

In heat pumps, this restriction is accomplished with two check valves and metering devices. The metering devices block the refrigerant flow into the evaporator, maintaining high pressure in the refrigerant line between the condenser and evaporator and allowing a small amount of refrigerant to pass into the evaporator and expand. Only one metering device restricts the refrigerant flow at any one time. The restricting device will be the one closer to the evaporator coil, and the evaporator coil, as previously explained, will change as the heat pump is switched from heating to cooling, and back again. As shown in Figs. 10-1 and 10-2, the check valve nearer the evaporator will be closed to force the refrigerant through the metering device. The check valve near the condenser will be open to allow the refrigerant to flow freely around the metering device.

One of the changes that occurs when a heat pump switches from cooling to heating is that the check valve at the inside coil opens (it was closed on the cooling cycle). The check valve at the outside coil closes (it was open on the cooling cycle).

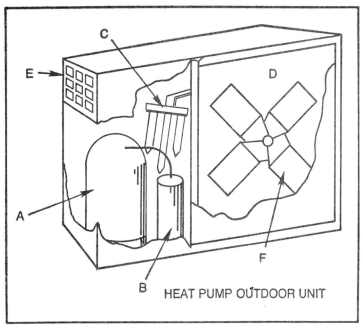

Fig. 10-3. The outdoor unit of a split-system heat pump. (A) is the compressor that pumps refrigerant through the system. (B) is the refrigerant accumulator that collects excess refrigerant inside the system. (C) is the reversing valve (D) represents the outdoor coil that is an evaporator on heating cycle and a condenser on cooling cycle. (E) are defrost relays (F) is the fan that blows air over the coil.

Reversing Valve

To change the heat pump from the cooling cycle to the heating cycle there must be some way to change the direction of refrigerant flow through the system and to switch the functions of the indoor and outdoor coils. The reversing valve serves this function. Notice in Figs. 10-1 and 10-2 that the refrigerant enters and leaves the compressor in the same direction for both cycles. But the reversing valve position is changed so that in the cooling cycle the refrigerant first enters the outdoor coil (the condenser), and on the heating cycle the reversing valve is positioned so that the refrigerant first enters the indoor coil (again, the condenser).

Defrost Cycle

The heat pump is equipped with an automatic defrosting system to remove ice and frost that builds up on the outdoor evaporator coil during the heating season when the temperatures drop below about 40°. If the frost were not removed, it would eventually accumulate to the point that there would be no air flow over the evaporator, and the refrigeration system would not work.

The defrosting process takes only a few minutes. You may see some steam coming from your heat pump's outdoor coil when it is defrosting, but this is caused only by the melting frost vaporizing in the cold air. When the defrost cycle ends, the steam will end, too.

In one type of defrosting system, hot gas is diverted from the compressor to the outdoor evaporator long enough to melt the frost from the coil. The outdoor fan turns off, and supplementary heaters supply necessary heating requirements for the house. The defrost cycle takes only a few minutes. Servicing the heat pump's defrost cycle is discussed later in this chapter.

Self-Contained Heat Pump Systems

The heat pump systems we have discussed so far have been the type that is commonly known as the split system heat pump. A heat pump of this type has two units—an indoor unit and an outdoor unit—connected by tubing that carries refrigerant from the outdoor coil to the indoor coil. In the split system, the compressor, the outdoor coil and most of the heat pump's relays are housed in a steel cabinet outdoors. The indoor coil, supplementary heaters and fan that blows the air through the duct system are housed indoors in the furnace room. The indoor unit merely is placed in the house where you would otherwise place a furnace, and the duct system connects to it in the same manner as a furnace would connect.

With a self-contained heat pump, all the components are the same, but all heat pump components—even the "indoor coil" and supplementary heaters—are housed in one single unit that sits outside the house. Figure 10-5 shows a self-contained heat pump. In this type of system, the duct system is brought through the foundation to connect to the heat pump. The unit's operation is the same as with a split system. The only difference is that room air is ducted out to the heat pump and circulated through a coil outside to be warmed or cooled, rather than being circulated through a coil and unit indoors.

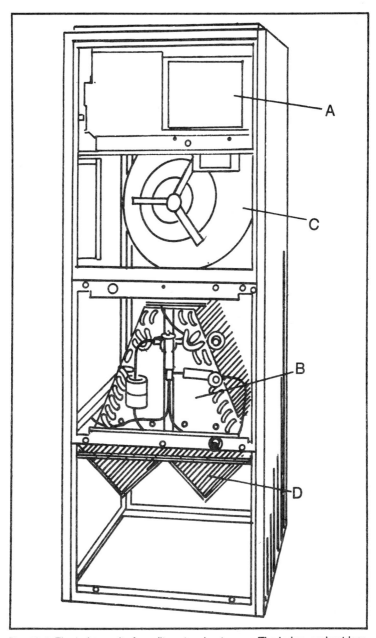

Fig. 10-4. The indoor unit of a split-system heat pump. The indoor and outdoor units are connected with refrigerant lines. (A) is the electric strip heater compartment. (B) is the indoor coil that is evaporator on cooling cycle and condenser on heating cycle. (C) is a squirrel-cage fan that blows air through indoor coil and duct system. (D) is the air filter (courtesy The Williamson Company).

SUPPLEMENTARY HEAT AND THE BALANCE POINT

As the outdoor temperatures drop, there is less heat in the air for the heat pump to pick up and transfer into the living area. At some temperature the heat pump's refrigeration system will not be able to adequately heat the house and supplementary heating elements will turn on to assist it. This temperature is known as the system's balance point, and the balance point varies according to heat pump model, size and installation. For many systems, the balance point will fall between 20° and 30°, but on some systems the balance point will be lower or higher. A low balance point, or course, means your system is more efficient and is relying to a larger extent on the less expensive refrigeration system heat. But there is another side to the coin—a lower balance point may mean a more expensive heat pump or a heat pump that is actually oversized. Thus, it might be cheaper to pay less for a system that will have a slightly higher balance point but which will be a little less energy efficient.

Most heat pumps use electric resistance heating elements to supply supplementary heat, but gas and oil-burning supplementary heaters have also been used with success. The electric resistance heating unit is divided into several elements, or strips, which are often called strip heaters. A typical heat pump might have four resistance elements. In better heat pump installations, each strip comes on independently and heats up only if the refrigeration system and prior strip heaters have been unable to supply needed heat. This saves wasting electrical energy that would be used to heat up all the strip heaters when a single heater used a little longer would have supplied the required heat.

When the strip heaters come on the heat pump is operating partly as a refrigeration unit and partly as an electric furnace. As the temperature outdoors continues to drop, the heat pump relies more and more on the electric resistance heaters to supply the needed heat for the home. Finally, the temperature will drop to a point where the refrigeration unit will shut down completely and only the electric heaters will supply heat. At this point, the heat pump is operating just like an electric furnace. In some systems, the compressor will continue to run with the strips.

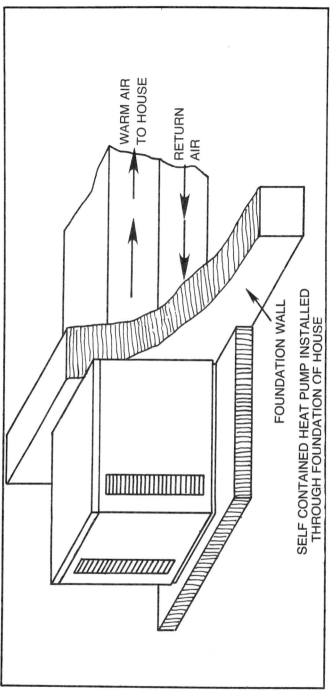

Fig. 10-5. A self-contained heat pump. The self-contained system has both coils housed outdoors in the same cabinet, and the duct system is brought out to it. In this unit, air from the living area enters the heat pump through the lower duct. Inside the cabinet, the air is ducted through what would be the "indoor" coil on the split system heat pump. This coil heats or cools the air, just as the indoor coil of a split system heat pump would. After circulating through this coil, the air is ducted back into the living area.

When the strip heaters come on, the heat pump's efficiency drops. For this reason, heat pumps are little more efficient in very cold winter temperatures than regular electric furnaces. However, one of the design improvements in heat pumps in recent years has been improving the refrigeration system's low-temperature performance so that heat pumps can operate at lower temperatures without turning on the supplementary heaters.

Outdoor Thermostats

Outdoor thermostats are used to keep the electric strip heaters from coming on until they are really needed. These thermostats are separate from the indoor thermostat. They sense the outside temperature and will allow the indoor thermostat to bring the strip heaters on only when the outdoor temperature is below a preset level. Thus, if your heat pump has four strip heaters, one strip heater can be set up to come on to assist the pump when a sudden gust of cold air drops the temperature more than 3° below the thermostat setting. This would assist the pump in bringing the room temperature quickly back to the desired setting and tends to remove fluctuating temperatures indoors on gusty days or when the family is making numerous trips outdoors. The second strip heater can be set with an outdoor thermostat to come on only if the outdoor temperature is below 35°. The third and fourth strips might be set for 25° and 15°. See Fig. 10-6.

If there is no outdoor thermostat, as soon as strip number one comes on, strip number two will begin energizing. Then, number three and number four will come on shortly. The wasted heat in such an arrangement is apparent if you imagine what would be happening on a 40° day when people are going into and out of the house frequently. The temperature would frequently drop more than 3° below the thermostat setting, and all four heaters would turn on to help the heat pump recover the indoor temperature quickly. All day long, all four heaters would be turning on and off!

With an outdoor thermostat, however, the next element in the sequence cannot turn on unless the outdoor thermostat senses that the outdoor temperature is below the setting at which the strip may be turned on. Thus, on a 30° day, the pump might not be able to supply all the heat the house

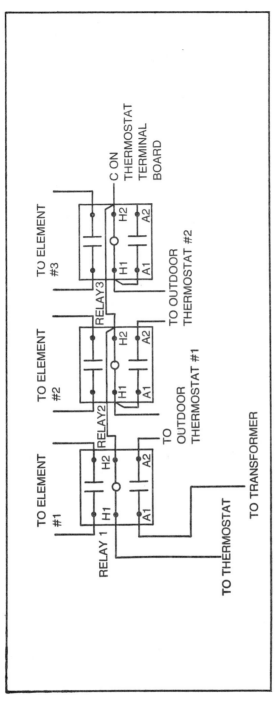

Fig. 10-6. This wiring diagram of the relays that control the electric resistance heating elements of a heat pump illustrates how the outdoor thermostats work. The outdoor thermostat interrupts the connection between A2 on element number one and H1 on element number two. Thus, unless the outdoor temperature is below the level set in outdoor thermostat number one, this connection will be open and element number two cannot come on. Outdoor thermostat number two installed between A2 and H1 of elements two and three will not allow element number three to come on with elements number one and number two unless the outdoor temperature is below the level set on outdoor thermostat number two.

On some installations there may be an outdoor thermostat installed between the thermostat and H1 of element number one. It is common, however, to install the system so that the indoor thermostat turns on element number one when the heat pump cannot recover the indoor heat quickly enough. Thus, as here, the first strip will come on when the indoor temperature drops 3° below the thermostat setting, regardless of outdoor temperature.

requires. When the thermostat senses that the indoor temperature is more than 3° below the thermostat setting, it energizes strip number one. Strip number two is then energized, and since the outdoor temperature is below the 35° outdoor thermostat setting, strip number two will come on. Strip number three would come on next, but since the outdoor temperature of 30° is above the 25° outdoor thermostat setting for number three, strip number three will not come on. Strips number one, number two and the pump will supply all the heat for the house.

Efficiency and Outdoor Thermostats

Not all heat pumps are installed with outdoor thermostats, but for the best efficiency and lowest operating cost, you should be sure your heat pump is. The extra cost is well worth it in electricity savings! You can install as many outdoor thermostats as your heat pump has strip heaters, but for most installations, three or four outdoor thermostats are sufficient. If you have more than four strips, two strips can be set up to come on at the same outdoor temperature.

Setting the temperature of these outdoor thermostats is something you should work out with your dealer. When you purchase your heat pump, be sure to tell him that your primary interest is in energy saving, and that you want the strips to turn on no sooner than necessary.

For the most efficient heat pump operation, the first strip would be set to come on at about the balance point, rather than at 3° below the thermostat setting. This would keep the strips off when you turn the thermostat up from 60° to 68°. Without the outdoor thermostat, the first strip would come on because the indoor thermostat would be at a 68° setting and the inside temperature would be 60°—more than a 3° temperature difference. In time, the pump alone would make up the 8° temperature difference if there is an outdoor thermostat to keep the first strip off. The disadvantage to keeping the strip off is that it may take the pump quite a while to bring the temperature of the house up those 8°. A heat pump does not generate the high air temperatures of most furnace systems. Supply air from a heat pump is about 100° to 125°, quite a bit cooler than the 125° + air temperatures that occur with other heating systems. Thus, the heat pump cannot recover the temperature of a building as quickly as other heating systems, and the occupants may feel cool for a while if the pump alone has to make up that 8° temperature difference.

Your heat pump dealer and installer is the one most familiar with your heat pump system and your home's heating needs. He is the one who is best able to adjust the outdoor thermostats according to the heat pump's design and its ability to supply required heat for your home. You should tell him that you prefer to have the outdoor thermostats set to save you the most energy possible, and then listen to his ideas on which arrangement will give you the best results. If you're willing to sacrifice a little comfort in the way of cooler room temperatures while the pump recovers the heat in the house, be sure to tell him. Usually the number one strip will operate with the compressor. When the system goes on defrost, the elements need to operate. The system could be wired with an outside thermostat on the number one strip, and then this outside thermostat can be bypassed on the defrost cycle.

Gas and Oil As Supplementary Heat Fuels

Although electric resistance heaters are the most common supplementary heating units on heat pumps, gas and oil-burning units are also available for supplementary heat. Many people prefer them to electric supplementary heaters because they usually supply more heat per dollar than electric resistance heat, and this means their supplementary heat costs less. So far, electricity has been the most popular supplementary heating fuel because heat pumps have been installed mostly in areas where natural gas or oil were not available or were quite expensive. But you can expect gas and oil to grow as supplementary fuels for heat pumps in areas where they are available. Their cost is more expensive than pump heat but less expensive than electrical resistance heat.

Supplementary heaters are optional on most heat pumps, but they are options you can't afford to be without unless you live in an area where you're confident the temperature will never drop below about 30°. One way to supply supplementary heat if you are thinking of switching your furnace for a heat pump is to retain the furnace for supplementary heat and add heat pump coils and lines to your existing furnace assembly. Chapter 6 shows how you can do this.

OTHER TYPES OF HEAT PUMPS

The heat pump systems described so far, whether split system or self-contained, can all be categorized as *air-to-air* heat pumps. This means simply that the heat pump transfers heat between outside air and room air inside the house.

Other types of heat pump systems have been developed, with varying success, that would transfer heat between underground *earth* or underground *water* and the room air. The former system is called a *ground-to-air* heat pump, and the latter is called a *water-to-air* heat pump. Ground-to-air heat pumps are available for home installations, but they have often experienced problems because the earth is not as good a conductor of heat as air. Water-to-air heat pumps are not generally available for residential use. The air-to-air heat pump, therefore, remains the standard heat pump for residential heating and cooling.

HEAT PUMP EFFICIENCY

The heat pump's high efficiency and relatively low cost of operation are the basic reasons behind the surge of interest in heat pumps, but they are also among the most misunderstood aspects of heat pumps. The heat pump is a year-round climate control system, supplying air conditioning as well as heating, but it is on the heating cycle that the heat pump has so much energy-saving potential. On the cooling cycle the heat pump is about as efficient as any other air conditioner. Each heat pump model's cooling efficiency is rated by an Energy Efficiency Ratio (EER), just like any other air conditioner. By comparing EERs on different heat pump models, you can determine which unit is most efficient on the cooling cycle.

Since it's the heating cycle of the heat pump that saves most of the energy, that's the one you should really concentrate on when deciding whether to purchase a heat pump. The job is made more difficult because heat pump efficiency varies according to general climate and outdoor temperature.

There are two figures commonly used in rating heat pump efficiency on the heating cycle. The Coefficient of Performance (COP) is the figure most commonly used, and it is the ratio of heat output divided by electricity input. An electric furnace has a COP of about one, and a heat pump's COP might average 1.2 to over three, depending on the climate and installation. The Seasonal Performance Factor (SPF) is a figure that tells the heat pump's efficiency during the entire heating season—thus it is a measure of the *average* COP during the entire heating season. Again, an electric furnace has a SPF of about one, and heat pump SPFs should range upward from that. The higher the COP and SPF figures, the more efficient the heat pump.

Heat pump efficiency, as mentioned, varies with the outdoor temperature. As the temperature drops, there is less heat in the air for the heat pump to absorb, and at some point the electric heating elements (COP of one) begin heating much of the air. It's easy to see that when the electric heating elements are on, the heat pump's overall efficiency drops.

For this reason, heat pumps have traditionally been more popular in the warmer winter climates of the South than they have been in the North. The heat pump is extremely efficient at temperatures above the 20-30° range.

In most parts of the United States, a heat pump will have an *average* heating efficiency of about 120 to 200 percent (SPF—1.2 to 2.0) over an entire heating season. Some areas in the southern quarter of the United States where very mild winters are the rule may have somewhat higher SPFs. This means that the heat pump will yield 1.2 to two times the amount of heat a fully efficient electric resistance heating unit would yield while using the same amount of electricity. Translating that efficiency use into savings, that means a heating bill savings of about one-third to one-half the cost of electric resistance heat.

The exact SPF in your area will vary with the heat pump's design, your local climate and the heat pump installation in your home. Each installation yields slightly different SPFs, even for the same model. As a general guideline to determine whether a heat pump would be cost-efficient for you, you should figure on the following SPFs, as shown in Fig. 10-7. If you live in the northern third of the United States and have relatively harsh winters, you can expect an SPF of about 1.25 to 1.5. In the middle third, or in areas where winters are usually moderate, the SPF will be about 1.5 to 1.75. In the southern third or in areas that have relatively mild winters, the SPF will be about 1.75 to two.

Some heat pump models will perform more efficiently with a higher SPF, but these figures give you the basis to make a conservative efficiency estimate of a heat pump's performance in your area and to decide whether a heat pump will be a cost-efficient heating system for you. You can also expect that in the next few years compressor designs will improve even more, and efficiency-increasing installation techniques will become more prevalent. These changes will further increase the average efficiency of heat pumps over the entire heating season. Even now, many new heat pump installations in your area may achieve SPFs that are considerably higher than these given. Actual test figures are difficult to come by, however, and you should remain skeptical of efficiency claims of 300 percent over an entire heating season.

MORE BUYING TIPS

So far, we have tried to give you some general guidelines to follow in deciding whether a heat pump will be a money-saving investment for your home, but they are only that—general guidelines. Each installation is different, and each homeowner can expect different results from his decision to install a heat pump. By necessity, we have been very general in our discussion. Before you can make an intelligent decision about a heat pump for any given home, you must get advice from experts familiar with your own climate and your home's heating requirements. Heat pump dealers (always talk with several, if possible), heating system contractors, building contractors, power company representatives and heat pump owners can give you a good idea of how heat pumps in your area perform.

Heat pump dealers and power company representatives are in the best position to help you decide what the efficiency of a heat pump would be in your area and whether a heat pump is the right heating system for your needs. Be sure to choose reputable dealers to help you with this decision. You want dealers who will not pressure you into buying a heat pump and who will not use expert knowledge to draw you into a sale that is not in your best interests. What you want is straight talk, not sales talk, and you should be able to get it from a reputable dealer.

Beware of exorbitant efficiency claims when discussing heat pumps with salesmen. It's possible to show an efficiency of 300 percent and more for a heat pump that actually on a

HEAT PUMP SPF GUIDE TO HELP ESTIMATE EXPECTED HEATING SEASON COST.	
REGION OR WINTER CLIMATE	SPF
NORTHERN THIRD OF U.S.; HARSH WINTERS	1.25-1.5
MIDDLE THIRD OF U.S.; MODERATE WINTERS	1.5-1.75
SOUTHERN THIRD OF U.S.; MILD WINTERS	1.75-2.0

Fig. 10-7. This table gives you a rough, general quide of the average COP (coefficient of performance) you can expect from a heat pump over an entire heating season. These SPF (seasonal performance factor) figures are actually conservative guidelines to help you decide whether a heat pump is likely to save you money on your heating bills. Many heat pump models will deliver more efficient performance than this table indicates. Since climate is more important than geographic location in using the table, match your winter climate to the climate that best fits your locality.

In some heat pump installations, SPFs significantly higher than these may be obtained. Factors that lead to higher SPFs are heat pump capacity and installation techniques. For a given house, a larger capacity heat pump will have a higher SPF, but if the unit is too large, it will be vastly oversized for the cooling load, causing summer cooling problems. A modern heat pump compressor design installed with a two-stage thermostat and outdoor thermostats will also lead to some higher SPFs.

seasonal average is only twice as efficient as an electric furnace—and it's the seasonal average you're interested in, not the peak laboratory performance.

Finally, consider your needs carefully. Any heating system is a 10-year investment, so treat it as such! Examine your present heating and cooling requirements, and your future ones. Look at the features of the different heating/cooling systems, and try to pick the one that's best for your home—not the type of system your neighbor likes best. You're the one who will have to live with the system—not your neighbor.

MAINTENANCE

A heat pump depends more than most heating systems on a sufficient air flow to insure proper efficiency and operation. For this reason, you should be even more conscientious about performing routine maintenance if you own a heat pump than if you own another type of heating system.

You should clean the air filter about four times a year. If you have a split system heat pump, the filters are likely to be in the indoor unit in locations similar to those that are common on a central furnace system. Popular locations are in the return air duct at the return air grill in the living area or just before the duct enters the cabinet. On a self-contained heat pump, the filter will probably be just inside the cabinet in the duct where the air enters the unit. Be sure to check the possibility that your unit may have more than one filter. Cleaning filters is discussed in Chapter 4. Many heat pumps have electronic air filters.

The outdoor coil is in a natural location to get clogged up with leaves and dirt, which cuts into the efficiency of the heat pump, especially during winter when proper air flow over the outdoor coil is vital. Keep the outdoor coil clean and free of debris. If dirt gets clogged between the fins, brush them clean with water and a stiff-bristled brush. **Do not use wire brushes, or a metal object such as a knife or screwdriver to clean a heat pump coil. You may puncture the coil and cause refrigerant leakage.** If there is a screen over the outdoor coil, remove it and brush it clean. Keep shrubs, bushes, and trees away from the outdoor coil.

Check the fans to be sure their blades are clean. Squirrel-caged fans are especially prone to collecting dirt and grime that reduces the fan's effectiveness. The fan belts should have no more than ½ inch flex when you push on them, and you should examine them for wear and cracks. If the fan motors do not have sealed bearings, there will be a small oil cup on each end of the motor housing near the shaft. You should fill this cup with a few drops of 30W oil every six months.

Be sure the defrost drain pan is free of debris, and water will drain freely from the pan. If not, defrosted water will back up in the pan and ice will form, preventing a proper defrost cycle.

Your thermostat may have an indicator light that tells you when the compressor is not operating during the heating cycle. This lets you know that your supplementary heaters, and not the pump, are supplying the heat for your house. Since using the supplementary heaters costs considerably more than running the pump, you should make it a point to check this indicator daily and get a repairman to look at your unit as soon as possible when it comes on. Sometimes you may be able to start the compressor by turning the thermostat to cooling for three to five minutes, and then switching it back to heating. If the compressor starts, the light will go off.

Thermostats

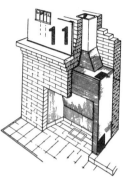

The thermostat is the heating system's master control—the device that tells the system when to come on, how long to stay on and when to shut off. All this, of course, is determined by the thermostat setting made by the homeowner. The homeowner sets this thermostat for the indoor temperature he desires, and the thermostat takes over from there. It turns on the furnace when the temperature drops below the homeowner's setting, and turns off the heat when the indoor temperature climbs above that setting. Everything is done automatically, and the homeowner need not lift another finger for weeks to keep his home at a comfortable temperature.

The thermostat controls the furnace by sensing the indoor temperature of the home being heated. A number of sensing systems are used for this purpose, but they are all quite similar. Thermostats generally rely on the expansion or movement of heat-sensitive metals to open or close electrical circuits that control the heating systems. See Fig. 11-1. The movement of these metal components opens and closes the contacts and turns the heating system off according to the temperature of the room where the thermostat is located.

TYPES OF THERMOSTATS

There are two basic types of thermostats: line voltage and low voltage. A line voltage thermostat (also called a high

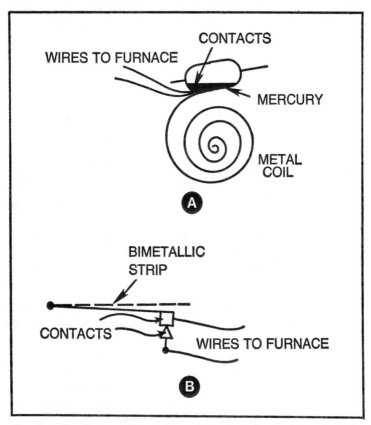

Fig. 11-1. Two common thermostat systems of sensing heat to activate the furnace. In (A), a bulb containing mercury is attached to a metal coil. As the temperature in the room falls, the metal in the coil contracts, tilting the bulb so the mercury flows to the bottom of the bulb where the two contacts are located. When the mercury surrounds these contacts the thermostat circuit is completed, energizing the heating relays to begin the furnace operation. When the room temperature rises, the heat expands the metal coil, tilting the mercury bulb forward to open the thermostat circuit and turn off the furnace. In (B), a bimetallic strip bends with changes in temperature to open and close the contacts. When the room temperature cools, the bimetallic strip bends so the contacts close to complete the thermostat circuit and turn on the furnace. As the room warms, the bimetallic strip bends back as shown by the dotted line, the contacts open, and the furnace turns off.

voltage thermostat) is wired directly into the power line going to the heating unit. This type of thermostat, shown in Fig. 11-2 and 11-3, is normally used with single-unit electric resistance heaters, such as baseboard heaters, ceiling cable and portable space heaters. Often the line voltage thermostat will be attached to the heater's cabinet rather than being installed remotely on a wall. A line voltage thermostat is 120 volts or 240 volts, depending on the voltage of the power line to the heater. Line voltage thermostats are discussed in Chapter 7.

A low voltage thermostat is the type most commonly used on furnace heating systems. In this type of system, shown in Fig. 11-4, the thermostat is not wired directly into the main power line, but instead is wired into a low-voltage line (usually 24 volts). A transformer in the furnace steps down the line voltage to 24 volts. When the thermostat contacts close, this completes the low voltage circuit. Power is fed to the electrical relays or gas valve and the furnace begins heating.

There are several different low voltage thermostat systems, but they can generally be grouped into two categories: single-stage thermostats and two-stage thermostats. The single-stage thermostat has one set of heating system contacts. When those contacts are closed, the entire heating

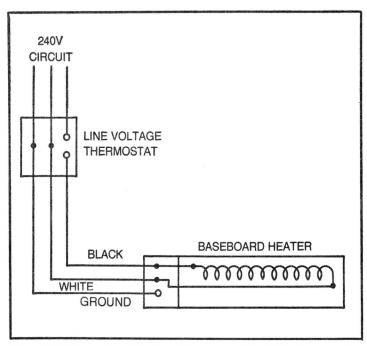

Fig. 11-2. The line voltage thermostat is connected directly into the power line going to the heating unit. In this case, the power line is 240 volts, so a 240 volt thermostat is used. The line voltage thermostat may be installed remotely from the heater on a wall, or it may be factory-wired into the heater and attached to the cabinet. In the latter case, when you connect the heater terminals to the power line you will also be wiring the thermostat into the electrical system.

Fig. 11-3. A line-voltage thermostat installed remotely from the heater on a wall. The cover of the thermostat has been removed to show the terminals (A) underneath. The power lines to the heater are connected to these terminals. The homeowner adjusts a dial to control the temperature of the room.

capacity of the furnace is energized. The two-stage thermostat, on the other hand, has two sets of heating system contacts. Each set of contacts turns on half the heating system's capacity, independent of the other. With a 2-stage thermostat, one set of contacts turns on half the furnace heaters. The second set of contacts does not turn on the second half of the heating capacity unless there is too much heating load for the first half of the heaters alone. Two-stage thermostats are used often as energy-saving devices for electric furnaces and heat pumps.

Single-Stage Thermostats

Single-stage heating-only thermostats are the simplest type of low voltage thermostats. This thermostat has a single set of contacts that closes when the temperature drops, and the closing contacts turn on the entire capacity of the heating system. A heating-only thermostat cannot be used with central air conditioning because air conditioning requires a thermostat with a second set of contacts that close when the temperature rises during the cooling season.

A single-stage heating-cooling thermostat is designed for central heating/air conditioning installations. It includes two sets of contacts. One set of contacts is connected to the heating system and closes when the temperature falls. A second set of contacts is connected to the cooling system and closes when the temperature rises (Fig. 11-5 and 11-6).

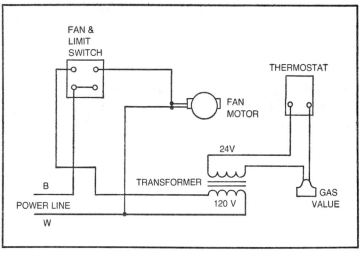

Fig. 11-4. A simplified wiring diagram for a furnace. A transformer in the furnace steps down the line voltage to 24 volts for the thermostat circuit. The thermostat opens and closes the circuit, but notice that the gas valve (on a gas furnace) is also wired into this circuit. Thus, when the thermostat closes the circuit, power is sent to the gas valve and this starts the furnace burning fuel. In an electric furnace or oil furnace an electric relay replaces the gas valve to activate the furnace heating system.

Fig. 11-5. A single-stage heating/cooling thermostat.

The single stage thermostats—whether heating-only thermostats or heating-cooling thermostats—are used primarily for gas furnaces and oil furnaces. On most of these furnaces in residential installations, all the furnace's burners turn on at once when the thermostat calls for heat. In other words, these furnaces are not "staged." Single-stage thermostats are also used with non-staged electric furnaces (that is, where all heating elements come on at once), but such installations are inefficient because of the amount of electricity that is wasted each time a cold electric element must be heated.

Two-Stage Thermostats

It is wasteful for all heating elements of an electric furnace to come on for a brief time to recover a slight temperature drop inside a building. It's much more efficient for half of those elements to come on for a longer time to do the same job, and that saves the wasted electricity that would otherwise be used heating all the elements.

The two-stage thermostat takes advantage of this energy-saving possibility. When the two-stage thermostat calls for heat, only about half of the furnace's heating elements

Fig. 11-6. Single-stage heating-cooling thermostat with cover removed. (A) mercury bulb contacts mounted on metal coil (B). Dial (C) switches thermostat and furnace system from heating to cooling.

are used—the first two elements of a four-element furnace, for example. Then, if the furnace cannot recover the temperature and the indoor temperature continues to drop, the second stage of the thermostat is energized and the second half of the heating elements are turned on.

When a two-stage thermostat has been installed, we say the furnace has been "staged." Two-stage thermostats are used as energy-saving devices on electric furnaces and on heat pumps, but not all of these heating systems are installed with two-stage thermostats. There are a number of reasons for this. Manufacturers often ship their products wired for the simplest thermostat installation—the single-stage thermostat. The installer may not want to take the time and effort to wire in the two-stage thermostat. The installer may also want to avoid the additional cost to his customer the two-stage thermostat would entail. Some furnaces use time delay sequences (about five to 10 seconds apart) to help hold the last elements off.

A two-stage heating-cooling thermostat is shown in Fig. 11-7 and 11-8. The two-stage thermostat has two mercury bulbs for heating and one for cooling. The first heating stage is activated when the indoor temperature drops about 1.5° below the thermostat setting. If the electric furnace is a 20 kilowatt furnace, comprised of four 5 kilowatt elements, the first stage would turn on elements one and two, or 10 kilowatts. On mild days, the first stage will probably be enough to heat the house, and the second stage of the furnace will never have to come on. In the same manner, if a heat pump is connected to a two-stage thermostat, the first stage brings on the pump, which will be enough to heat the house on mild days.

On cooler days, however, the temperature inside the house will continue to drop because elements one and two or the pump are insufficient to supply the entire house's heating needs. When the temperature drops to about 3° below the thermostat setting, the second heating stage will be activated, and the elements three and four will come on. On a heat pump, the second stage turns on the supplementary heat.

OUTDOOR THERMOSTATS

For even more efficient use of the electric heating elements, a two-stage thermostat can be combined with an outdoor thermostat to keep the second heating stage from turning on unless it's really needed. An outdoor thermostat is a control that senses the temperature outdoors and will not allow the second stage to turn on unless the outdoor temperature is below a preset level, usually between about 20° and 35°. This prevents inefficient use of the second stage of the heating system when the thermostat is turned up suddenly from 60° to 68° or when a sudden gust of wind lowers the room temperature. Without an outdoor thermostat, both stage of the two-stage thermostat would be activated and all four heating elements would turn on, even though the weather might be mild enough that two elements would be sufficient.

On a heat pump, it is even more important that an outdoor thermostat be installed to keep from bringing on the stage two supplementary heaters. With an electric furnace there will be a noticeable, but small, loss in efficiency if both stages are activated when stage one is sufficient to take care of the home's heating needs. But a heat pump has a large efficiency difference between stage one and stage two. Stage one is the efficient pump, while stage two is the comparatively inefficient

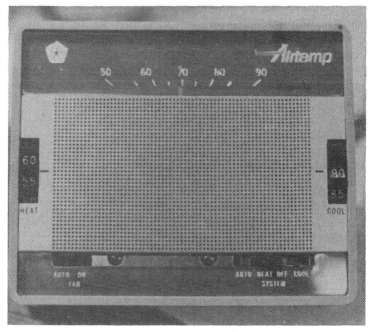

Fig. 11-7. A two-stage thermostat with cover.

and expensive supplementary electric strip heaters. Naturally, you want to install the heat pump thermostat to keep the strip heaters off unless they are really needed, and even then you only want to turn on as many strip heaters as necessary. Outdoor thermostats do this.

In recent years, some heat pump installers have begun connecting a separate outdoor thermostat for each electric heating element. Thus, the first strip can be set to come on at temperatures below 35° or when the heat pump is on defrost cycle, the second at temperatures below 25°, the third at 20° and the fourth at 15°. The exact temperatures at which outdoor thermostats are set depends on the size and capabilities of the heat pump, as well as the size of house and particular installation. Outdoor thermostats are discussed in connection with heat pumps in Chapter 10.

With an electric furnace, only one outdoor thermostat is usually required. Unlike a heat pump where large efficiency sacrifices are being made each time an additional electric heating element turns on, an electric furnace does not sacrifice a great deal of efficiency when additional elements are operating. And bringing on the additional heating elements helps recover the heat in the house much faster than leaving that chore to part of the heating capacity.

Usually the outdoor thermostat is installed as part of the electrical circuit for the first heating relay that is activated by the second stage of the indoor thermostat. For example, you would install an outdoor thermostat in the electric circuit for element three of a four-element electric furnace, or in the circuit for element four of a six-element electric furnace. The installation of the outdoor thermostat keeps the second half of the elements turned off unless the outdoor temperature is below the preset temperature.

The outdoor thermostat operates a bit differently than an indoor thermostat in that it does not turn anything on itself. The indoor thermostat remains the furnace's main control. All the outdoor thermostat does is keep the electrical circuit for some elements open so those elements cannot turn on even when the indoor thermostat calls for heat.

The outdoor thermostat has a set of contacts that close to complete the electrical circuit once the outdoor temperature drops below a pre-set level. But once the outdoor thermostat closes the circuit, the indoor thermostat must call for heat and turn on that circuit before the elements will begin heating.

THERMOSTAT MAINTENANCE

A thermostat is a sensitive electronic instrument that seldom needs major adjustment. In fact, a person with no knowledge of thermostats who begins prying indiscriminately on his unit is sure to do much more harm than good. The fact is that beyond a few simple routine checks, there's not much you can do to make your thermostat work better. But if you aren't careful, you can certainly throw the entire unit out of whack by bending sensitive metal springs and coils or by messing with the calibration systems.

To insure proper thermostat operation, there are two basic things the homeowner can do. Be sure the thermostat is in a good location, and be sure the thermostat is level when installed. These two things will do more than anything else to make the thermostat control the furnace properly. Thermostat location is discussed in the next section.

Although all thermostats should be installed level, mercury bulb thermostats are particularly sensitive to mistakes in

Fig. 11-8. Two-stage thermostat without cover. (A) the two heating system mercury bulb contacts. (B) The single cooling system mercury bulb contact. (C) Row of terminal connections that are wired to the furnace terminal board to connect thermostat.

leveling the thermostat—for obvious reasons. Of course, a thermostat may be level when installed and then become unlevel after months and years of use. A hard knock may jar the thermostat loose, or time can make the mounting screws pull out.

To be sure the thermostat is level, turn off the power to the furnace and carefully remove the faceplate from the thermostat. Once you have the cover removed, proceed carefully to avoid improper bending of coils, strips and contacts. The thermostat base is usually plastic, and somewhere on the base you should find a line marked "level" or two prongs extending from the base suitable for resting a level on. With a round thermostat, you may have to remove the entire thermostat mechanism from the base to level it.

Place a small level on the thermostat base, and adjust the base's position in relation to the wall as needed to make the base level. Also check the wall on which the thermostat is mounted to be sure the thermostat is mounted perpendicular to the floor. If the thermostat is not perpendicular but leans inward or outward a bit, now is a good time to shim up the top or bottom of the thermostat base as needed to make it perpendicular.

The internal parts of the thermostat should be lightly dusted occasionally to keep dust from accumulating on the contacts or other thermostat parts. This should be performed lightly with a soft brush. Be careful not to bend the bimetal switch or coils!

It's a good idea to check the room temperature indicator on the thermostat to see if it's indicating the proper temperatures, since the thermostat frequently gets adjusted according to the temperature that's indicated on the thermostat thermometer. The problem is this thermometer is frequently out of adjustment. Usually the thermometer is located in the thermostat cover and can be easily adjusted. **Do not attempt to recalibrate the dial or pointer that shows the thermostat setting. Adjust only the thermometer showing the actual temperature of the room (Fig. 11-9).** The thermostat indicator itself is rarely out of calibration.

To adjust the thermometer, you'll need a reliable room thermometer—not a glass bulb attached to a cardboard backing with calibrations. Leave the room thermometer and the

Fig. 11-9. If you feel you should adjust the thermostat thermometer to indicate the correct room temperature, be sure to adjust the correct pointer. On this thermostat, (A) is the pointer to recalibrate. This is the indicator that shows the room temperature. (B) is the indicator that shows the thermostat setting selected by the homeowner, and this indicator should not be recalibrated.

thermometer from the thermostat together in the same location for about 10 minutes. Compare the two thermometers, and adjust the thermostat thermometer as needed to make it match the room thermometer reading. Of course, probably the best way to handle any problems that might be created by an incorrect thermostat thermometer is to simply not pay attention to that thermometer. Set the thermostat to the temperature desired, and leave it. Adjust it upward or downward according to whether the room is comfortable—not according to the temperature reading on the thermostat thermometer.

Fig. 11-10. This type of thermostat anticipator has the anticipator adjustment located on dial (A) in front of the bimetal coil that senses the heat in the room and operates the mercury bulb contacts. To adjust the anticipator to match the amperage of your furnace's thermostat circuit, move the pointer (B).

THERMOSTAT ANTICIPATOR

Most thermostats have anticipator circuits to help the thermostat keep the room temperature steady—within a couple of degrees of the thermostat setting. The anticipator is actually nothing more than a resistor wired into the thermostat, and this resistor produces heat that warms the thermostat about 1° above the room temperature. Some of these anticipators are adjustable, and some are fixed, but in any case the anticipator current capacity should equal the current flowing through the low-voltage thermostat circuit.

One purpose of the anticipator is to prevent furnace *overshoot*. If the thermostat is set at 68°, the anticipator will cause the thermostat to turn the furnace off when the room temperature is 67°. When the thermostat cuts the furnace burners or heating elements off, however, there is still heat left in the heat chamber. The fan will continue running until this heat is removed, and this extra heat will usually be enough to bring the house temperature up to 68° as desired. Without an anticipator, this extra heat would push the house temperatures up to 69° or so and would cause wider variations in the house temperature during cycles.

Any time you install a new thermostat or examine an existing thermostat for proper operation, you should check the anticipator circuit to see that it is set properly. Figures 11-10 and 11-11 show two types of adjustable anticipator circuits.

Fig. 11-11. The anticipator circuit of this two-stage thermostat is adjusted by moving a pointer along a sliding scale (A) to make the anticipator match the amperage draw of the furnace's thermostat circuit. This amperage is marked on the gas valve, the furnace terminal board, or in the owner's manual under "low voltage circuit current" or "thermostat circuit current".

The anticipator has a dial with numbers from .1 to 1 and a pointer (Fig. 11-12). The numbers on the dial represent the amperage of the low-voltage thermostat circuit, and the pointer should be set to the number corresponding to the amperage rating of that circuit for your furnace. The thermostat circuit amperage rating should be marked on the gas valve or on the furnace terminal board where the thermostat wires are connected to the furnace. If the amperage rating of your thermostat circuit is .2 amps, for instance, you should set the anticipator to .2. You can also find the amperage of the thermostat circuit with an ammeter.

You will also usually find the words "longer" or "shorter" on the anticipator adjustment. When the pointer is moved to the "longer" end of the scale, the furnace cycle is a bit longer, the furnace does not shut down quite as quickly, and the room temperature will be a bit warmer when the furnace runs.

THERMOSTAT LOCATION

Where the thermostat is located can have a large effect on how it performs. If it is located in a particularly cool room, the remainder of the house will be too warm. If it is located where cold gusts of wind frequently hit it, the same result will occur. Located near a window in the path of direct sunlight, the thermostat becomes too warm and the remainder of the house too cool.

The purpose of the thermostat is to sense the general temperature of the entire home—not the temperature of one isolated spot. To accomplish this, the thermostat should be located someplace that has good air circulation and that does not receive distorted temperatures. These are places your thermostat should not be located: behind draperies, in the path of cool winter drafts, near heat sources such as appliances, television, lights and heating ducts. The thermostat should always be located on a partition wall—never an outside perimeter wall. Because the outside wall borders on the cold outdoors during the winters, even the indoor side of a perimeter wall can be a few degrees cooler than the average temperature inside the house. Installing the thermostat on a cool wall makes it call for heat more often and raises fuel bills.

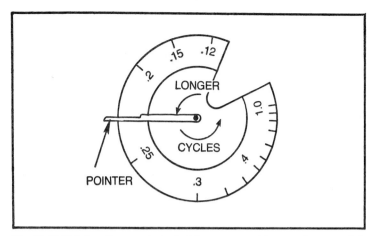

Fig. 11-12. This is a diagram of a typical thermostat anticipator adjustment dial. You should set the pointer to the amperage rating of your thermostat circuit. Moving the pointer in the direction marked "Longer cycles" makes the furnace cycle a bit longer.

If your thermostat is located in any of these unsatisfactory locations, it's probably costing you extra money in heating and cooling bills to keep it there, so you might as well consider moving it. Actually, moving a thermostat is a fairly easy job. You need to turn off the furnace power, remove the circuit wires from the thermostat terminals, and take the base off the wall. Be sure to tag the wires and terminals so you'll know which wires go on which terminals when you reinstall the thermostat.

Installing the thermostat in its new location is as simple as installing an electrical box. The low-voltage line is run down through the studs to the hole you've drilled in the wall. Attach the thermostat base to the wall and level it as described in the previous section. **Be sure the base is installed level.** Mercury bulb thermostats are particularly sensitive to mistakes in leveling. If the wall is not quite perpendicular, install shims behind the base at the top or bottom to straighten it up. All that's left to do is to connect the terminals to the wires, install the thermostat cover, and patch the holes made where you removed the thermostat.

The best place to install the thermostat is about 5 feet up on a wall in a central room, such as a living room. A hall often also makes a good place to install the thermostat as long as there is adequate air circulation through the hall. Avoid installing the thermostat in kitchens or in rooms where it would be unduly influenced by things such as fireplaces or wood stoves. On the other hand, if your family spends most of its winter time in the living room where the fireplace is located, you might want to install the thermostat there. Install it on an inside wall as far away from the fireplace or wood stove as possible, however. Also, you can expect that the other rooms in the house will be somewhat cooler due to the extra heat in the room with the thermostat. Do not install the thermostat where it will be directly in the line of any cold winter drafts. This would waste energy dollars because the cold drafts would bring the furnace on prematurely (Fig. 11-13).

If you are going to move your thermostat, you might consider installing an automatic set-back thermostat described later in this chapter. These devices allow you to set your thermostat to turn down automatically when the house is not in use to save energy and money. If you are moving your thermostat, this would be a good time to install one. Also consider installing a two-stage thermostat if you have an electric furnace wired into a single-stage thermostat. Replacing a conventional thermostat with either a set-back thermostat or a two-stage thermostat (if you have an electric furnace) will save you money in the long run.

THERMOSTAT SET-BACK

For years, some people have thought they might be able to save some money on their heating bill if they would turn down their furnace at night when the family is sleeping or during the day when the house is empty. After all, the reasoning went, you don't need the house to be as warm at night when everyone is sleeping. Adding blankets to the beds wards off any chill that would come from lower house temperatures. And if everyone is gone from home during the day, why have the furnace going full blast? If the furnace is turned down 10°' no one would even know the difference until he came home.

But there has also been an old wives' tale that accompanied that reasoning and prevented most people from carrying out their ideas of saving energy by turning down the thermostat. According to the old wives' tale, turning down the thermostat didn't really save any energy at all. When the thermostat was turned back up the furnace had to work so

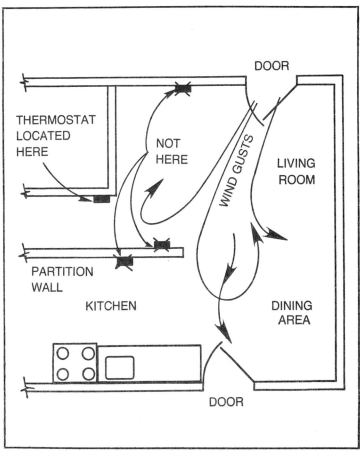

Fig. 11-13. When installing a thermostat or inspecting the location of an old one, avoid outside perimeter walls, kitchens, and areas likely to be struck by wind gusts. Place the thermostat on an inside partition wall that will block wind gusts.

hard to bring the house's temperature back up that it wasted more energy than was saved.

Old myths die hard, but statistics from the U. S Department of Energy have finally put this one to rest. It is simply untrue that you won't save energy by turning down your thermostat at night and then turning it back during the morning. According to the Department of Energy, if you set your thermostat back 10° for an eight-hour stretch each day during the heating season, you can save from nine to 15 percent on your heating bill (Fig. 11-14). It's apparent that setting back the thermostat is a worthwhile habit to get into! Likewise, during the summer cooling season, turning the thermostat up for several hours at a stretch will save you energy and money.

The cost savings can be significant. If you live in Cleveland, Ohio, and have a heating bill of $400, you will save about $48 a year ($400 × 12 percent) by turning down your thermostat from 65° to 55° each night. And chances are that except for the first few minutes each morning when the thermostat is turned up again, you'll scarcely notice the change.

The easiest and least expensive way to set back the thermostat is to do it manually. But this method raises objections with some homeowners, with the consequence that the thermostat doesn't get set back at all. Some people object to the fact that the house is cold for the first 15 minutes to half-hour after the thermostat is turned up. If the thermostat is set back and turned up manually, little can be done about this problem except to tolerate it. Another problem is forgetfulness. It's easy to forget to turn down the thermostat before going to bed at night or leaving the house in the morning.

Both of these problems are solved with automatic setback thermostats. These thermostats include a new component as part of the thermostat—a clock or timer—that allows the homeowner to set the thermostat to turn down the house temperature after bedtime and turn up the temperature half an hour or so before morning wake-up. This means you can now have your cake and eat it, too! The automatic thermostat will turn down the house temperature at the preset hour to save you money during the night or day when high room temperatures are not needed, and it will turn up the temperature before you wake up or come home so that the house will be warm and cozy.

There are a number of these automatic thermostats being sold nowadays, and most are being sold with do-it-yourself installation instructions. Instructions should be complete and easy to follow. Check the instructions before you buy. If you cannot follow them, tell the salesman to show you another model. Automatic thermostats are available at heating supply houses, hardware stores and even department stores.

There are three different devices available for automatic set-back. One type of device is a converter unit. The converter attaches to your present thermostat, and the timer device that comes with the converter automatically changes the

CITY	5°F	10°F
ATLANTA, GA	11	15
BOSTON, MA	7	11
BUFFALO, NY	6	10
CHICAGO, IL	7	11
CINCINNATI, OH	8	12
CLEVELAND, OH	8	12
COLUMBUS, OH	7	11
DALLAS, TX	11	15
DENVER, CO	7	11
DES MOINES, IA	7	11
DETROIT, MI	7	11
KANSAS CITY, MO	8	12
LOS ANGELES, CA	12	16
LOUISVILLE, KY	9	13
MADISON, WI	5	9
MIAMI, FL	12	18
MILWAUKEE, WI	6	10
MINNEAPOLIS, MN	5	9
NEW YORK, NY	8	12
OMAHA, NE	7	11
PHILADELPHIA, PA	8	12
PITTSBURGH, PA	7	11
PORTLAND, OR	9	13
SALT LAKE CITY, UT	7	11
SAN FRANCISCO, CA	10	14
SEATTLE, WA	8	12
ST. LOUIS, MO	8	12
SYRACUSE, NY	7	11
WASHINGTON, DC	9	13

Fig. 11-14. Use this table to figure your average savings if you turn your thermostat down during the night and/or day. These savings are based on a basic room temperature of 65°. Figures given are the approximate percentage saved with an eight-hour nighttime setback of 5°F and 10°F. (courtesy U.S. Department of Energy).

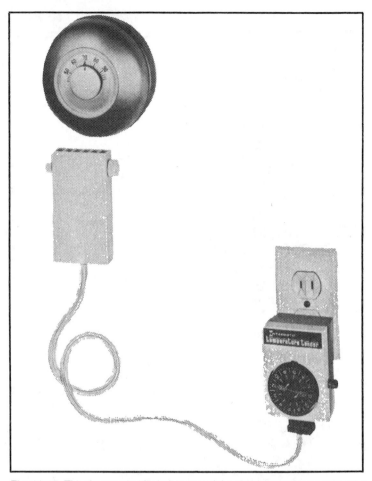

Fig. 11-15. This thermostat "fooler" is one of the simplest and least expensive ways to install automatic set-back thermostat. The timer plugs into a wall outlet. This timer is connected to a small heater unit which is installed under the thermostat. The timer causes the heater to generate heat, which "fools" the thermostat and causes it to sense the room is warmer than it actually is (courtesy Intermatic, Inc.).

thermostat setting according to the settings the homeowner has made. A second device, shown in Fig. 11-15, is a sort of thermostat *fooler*. This timed device is connected to an ordinary wall outlet. A small heater unit attached to the timer is installed on the wall just below the thermostat. This heater generates enough heat to warm the thermostat and make the thermostat "think" the temperature in the living area is normal. Actually, however, the room temperature is falling because the thermostat is not calling for heat as often. The timer automatically turns the "fooler" on and off at the time the homeowner desires.

The third way to achieve automatic set-back is to replace your present thermostat with a new automatic set-back thermostat, also called a clock thermostat (Fig. 11-16). This method gives you more options as to the type of thermostat and control system you can select, but it is also usually more expensive than other set-back devices. A new thermostat can cost from $50 to $100, depending on the type of thermostat and the control features you select.

Automatic set-back thermostats are available in single-stage, two-stage, heating-only and heating-cooling types. You can get them with a myriad of timing controls, but you can expect to pay more for a more sophisticated control system. Some thermostats allow three set-back periods daily and different set-back periods for each day of the week. If you are thinking of moving your thermostat anyway, it may make sense to consider installing an automatic set-back thermostat when you move it.

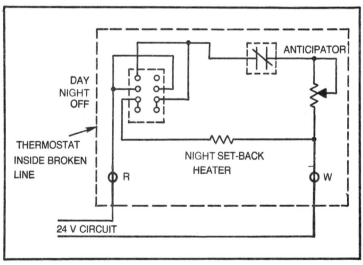

Fig. 11-16. A wiring diagram of a set-back thermostat. Notice that the set-back mechanism is actually a small resistor placed in the thermostat circuit. When current flows through this resistor during set-back hours, heat is generated. This heat "fools" the heat sensitive thermostat contacts and keeps them open, which allows room temperatures to drop a few degrees.

GUIDELINES

- Install a two-stage thermostat with an electric furnace or heat pump.
- Install one or more outdoor thermostats in the relay circuits of an electric furnace or heat pump.
- Set back your thermostat 10° at night or when the house is empty during the day. Consider installing an automatic setback thermostat so this won't have to be done manually.
- Check thermostat for proper location.
- When installing a thermostat, install on inside wall away from sources of heat or cold that might influence thermostat functioning.
- When installing thermostat, and as part of routine maintenance, set anticipator circuit for the amperage of the thermostat circuit.
- When installing thermostat, and as part of routine maintenance, check to be sure thermostat is level and perpendicular.

Humidifiers

It is common knowledge that humidity—or the amount of moisture in the air—affects comfort. Anyone who lives in a climate where the humidity varies at times, or who has traveled to high and low-humidity climates, knows how true this is. Even if the temperature outdoors is 90°, that high temperature doesn't feel as hot when the humidity is low as it feels when the humidity is much higher. That's because at low humidity the air contains a large capacity for absorbing moisture and will absorb perspiration from the skin quickly, cooling the body. But at high humidity the air is nearly saturated. It feels sultry and close, and perspiration evaporates quite slowly. Consequently, the body's built-in cooling system—perspiration—does not cool the body nearly as well and you feel much warmer.

If you think about it a bit, you can see how humidity relates to your home's heating system. If the humidity in your house is high (relatively speaking), you will feel quite a bit warmer than if the humidity in your house is lower. Thus, by raising the humidity of the air inside the house you can lower the thermostat setting on your furnace several degrees without sacrificing a bit of comfort for the occupants.

To increase the humidity in your house you can install a device called a humidifier. One type of humidifier is shown in Fig. 12-1. These devices can be attached to the furnace or duct system so they put moisture right into the furnace air as it circulates through the duct system. Another type of humidifier is a portable humidifier that is a self-contained unit inside a cabinet. The portable humidifier can be placed in any convenient location inside the home.

A lot has been written about the energy-saving potential of humidifier installation, and some of it is conflicting. The fuel savings aspect of humidifiers is discussed a bit later in this chapter. It is sufficient to point out here that in many instances, humidifiers can save money on fuel costs—enough savings, in fact, to pay for the humidifier within two or three years. But in other cases, humidifiers can bring minimal fuel savings or even net losses. Whether a humidifier will save you fuel money depends almost entirely on your home and how tightly it is constructed. If your home has few air leaks and has vapor barriers installed along the inside walls, a humidifier should save you quite a bit of money.

HUMIDITY BASICS

Fuel savings are only one reason to install a humidifier in a home, and the potential for saving fuel should not be the only reason you decide to install one. There are many additional benefits to humidifiers, so that even if your home is not constructed tightly enough to save much on fuel bills by adding a humidifier, you should consider adding one anyway.

As already noted, the humidity of a building affects the comfort of the occupants. In the winter months it is quite easy for the humidity inside the home to drop to a very low level—too low for comfort, really. Not only does this lower humidity take higher thermostat settings to compensate for the rapid evaporation of moisture from the occupant's skin, but overly dry air causes other unwanted problems. The skin feels extremely dry and may chap. Often a person's nose and throat will also become quite dry, which may make him more susceptible to germs and viruses that are a big enough problem during winter, anyway.

When humidity is very low, furniture can dry out and begin cracking. Dust inside the home is worse because carpets and fibrous products deteriorate faster and form microscopic particles that are broken off and fly about the house as dust particles. Electrostatic charges are much more prevalent

Fig. 12-1. This humidifier is installed in the furnace duct system to automatically put moisture into the house's air. The clear plastic pan holds water, and this water is picked up by the bristles of the brush rotating in the pan. The bristles carry the water into the air stream, the air absorbs some of the water, and the relative humidity of the household air is increased.

and become bothersome to the occupants. Merely walking across a carpet will build up enough electrostatic charge to give you a noticeable shock when you touch a metal doorknob. Increasing the amount of moisture in the house's air will go a long way toward eliminating most of these problems.

Humidity is often spoken of in terms of *relative humidity*. Relative humidity is a term you hear almost every time you listen to a weather report. The announcer speaks of the relative humidity is such-and-such percent and does not say there are this-and-that many grains of moisture in the air.

Relative humidity is the amount of moisture the air is holding relative to the amount of moisture the air is capable of holding. Since the air's capacity to hold moisture increases with the temperature, if the temperature goes up and the amount of moisture in the air remains the same, the relative humidity will go down. Take, for instance, air that has a relative humidity of 50 percent. This means that the air is holding one-half as much moisture as it is capable of holding at that temperature. The air has the capacity to absorb an additional quantity of moisture equal to the amount it already holds. Now suppose the amount of moisture present in the air remains the same, but the temperature of the air increases. At higher temperatures, the air has a larger capacity for absorbing moisture. So the relative humidity of the air will decrease even though the total amount of moisture in the air remains unchanged. Thus, the temperature rise might have reduced the relative humidity to, say, 30 percent.

It is easy to see how the relative humidity of the air inside a house gets so low in the winter, but not in the summer. The outside air is cold and has a small moisture-holding capacity. Thus, even if the relative humidity of the outside air is 50 percent, once that air is warmed to room temperature the moisture-holding capacity increases greatly. At room temperature, the relative humidity of the same air with the same amount of moisture might be as low as five or 10 percent—an uncomfortably low relative humidity.

Adding a humidifier to your home will add the needed moisture to the house's air during the heating season. Of course, during the summer cooling season the object is to reduce humidity in the air for comfort, and dehumidifiers do this. Air conditioners also remove humidity from the air.

Although the proper level of humidity will vary according to the preferences of the household, usually about 30 percent to 70 percent relative humidity is the most comfortable. To reduce static electricity charges, however, you may have to get the relative humidity up to about 50 percent. In very cold weather you probably will not be able to keep the humidity as high as even 35 percent relative humidity. Twenty to 25 percent may be all that's obtainable with a humidifier of an average capacity.

Fig. 12-2. The humidistat controls the operation of the humidifier according to whether the humidistat senses the relative humidity in the household air is at the homeowner's preset level.

For many humidifiers, adjusting the amount of moisture in the air is a simple matter of adjusting the *humidistat* to the desired relative humidity setting. The humidistat senses the amount of moisture in the air and turns the humidifier on and off according to whether it senses more humidity is needed (Fig. 12-2).

While having the relative humidity too low can cause problems in the home, having it too high also creates undesirable effects. One of the worst is peeling paint, caused when moisture-laden air comes in contact with a cold wall surface. The best way to prevent this problem is to have good wall insulation and have a vapor barrier installed on the inside wall. It is almost impossible to prevent some moisture condensation on window surfaces, however, because those surfaces are nearly always cold (Fig. 12-3). Moisture-laden air coming in contact with them will condense and form water droplets. Adding storm windows will cut out most condensation, reducing it to an amount that shouldn't be a problem in most cases. Double-pane windows will reduce condensation even further. Examine your window casings, and if excess condensation is causing the casing to deteriorate, reduce the relative humidity inside the house.

Excess moisture in the air can cause other problems, such as mildewing carpets and wood deterioration, but you shouldn't have these problems if you keep the relative humidity in the 50 to 60 percent range.

There are several methods of measuring relative humidity, but one of the easiest is a chart like the one shown in Fig. 12-4 that relates dry-bulb temperatures and wet-bulb temperatures to relative humidity. The dry-bulb temperature is measured with an ordinary thermometer. The wet-bulb temperature is measured with a wet-bulb thermometer—a thermometer that accounts for the effects of evaporation upon temperature and comfort. The wet-bulb thermometer has a dampened wick that covers the bulb. As the water from the wick evaporates, it cools the bulb and the thermometer's reading is lower. By examining the chart, you can see what a difference increasing relatively humidity makes on the apparent temperature (the wet-bulb temperature). Also notice that at 100 percent humidity, the dry-bulb temperature and the wet-bulb temperature are identical because no evaporation takes place.

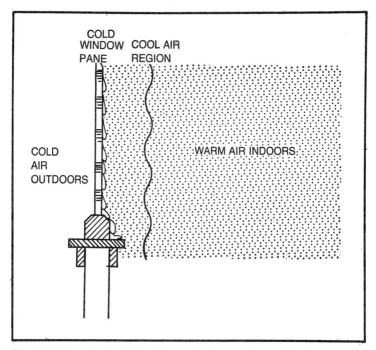

Fig. 12-3. When the humidifier is turned on the warm household air contains a good deal of water vapor, shown by the dots in the diagram. When this moisture-laden air strikes a cold surface, such as a window pane, the water vapor in the air condenses to form water droplets on the pane. This water can cause the wooden window casing to deteriorate if the water condensation is significant.

HUMIDIFIERS AND FUEL SAVINGS

At the first glance, it would seem like a snap to save money on fuel bills by installing a humidifier. As noted earlier, when the humidity inside a house is increased the occupants feel warmer and the thermostat can be turned down to a lower furnace temperature without any sacrifice in comfort. As a general guide, you can figure that for each 10 percent increase in relative humidity, you can lower the furnace setting by 1°. This means that if your furnace is now set for 72° indoor temperature and the relative humidity is 10 percent, by increasing the relative humidity to 50 percent you can turn the furnace down to 68°. The occupants will still feel as warm as they did before the change.

But this presents only one aspect of what happens when you add a humidifier. The other part is this: to increase the relative humidity of the home, the furnace air must evaporate water from the humidifier. This evaporation absorbs heat energy—energy that must be supplied by the furnace air—and this energy loss subtracts from the energy saved by adding a humidifier. Therefore, any energy savings accomplished by adding a humidifier must be net savings. That is, the amount of energy saved by reducing the thermostat setting must be greater than the amount of heat energy expended in evaporating the water to increase the humidity.

It takes more than 8000 BTU of heat energy to evaporate one gallon of water, and in the average home a humidifier will use about one gallon of water each day for each room in the house. Those are the energy costs of operating a humidifier, excluding the nominal amount of electricity needed to run the humidifier. As for energy savings, you will save about 3 percent on your heat bill for each degree you lower your thermostat. If you can turn your thermostat down from 72° to 68°, that's a 12 percent fuel savings.

In the average home, all these heat loss and heat savings figures usually add up to net fuel savings because the amount of heat lost through evaporation of the water is less than the 12 percent fuel savings. The net savings won't be 12 percent of course, but there will be a noticeable reduction in fuel usage.

However, this is not always the case in all homes. A poorly sealed home constructed with no vapor barriers and with numerous cracks that allow a good deal of infiltration air, loses moisture much faster than the average home. The more infiltration air your home has, the more water the humidifier has to evaporate to keep the relative humidity up inside the home. Thus, many homes constructed before the 1940's are poorly sealed, have no storm doors or windows, and may require two or more gallons of water a day per room. In these homes, the heat loss from evaporation is usually so great that it is higher than the fuel savings realized by turning down the thermostat. Of course, in such cases there is a net energy loss incurred from humidifier operation.

On the other hand, many modern homes are very well sealed and require much less water evaporation than average. In these homes the fuel savings from humidifier operation are even greater, and the fuel bill savings will pay for the humidifier in a year or two. A well-insulated home with vapor

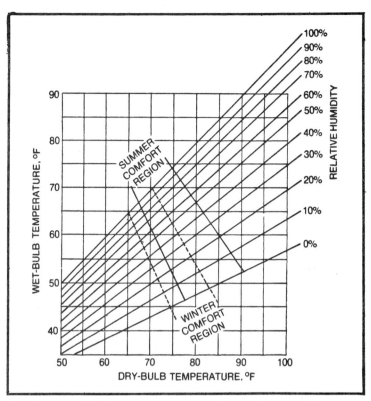

Fig. 12-4. A graph has long been used to determine the relative humidity by using the temperatures obtained from dry-bulb and wet-bulb thermometers. The relative humidity affects the apparent warmth of the air. This effect is indicated on the graph by the two comfort regions for summer and winter. Of course, the apparent temperature of the air is only one comfort factor. Most people also prefer the relative humidity to be between 30 and 70 percent for maximum comfort.

barriers, weatherstripping, storm windows and storm doors may require as little as one-half gallon of water a day per room. See Fig. 12-5.

Although precise statements as to the energy savings potential of humidifier installations depend on each house, we have tried here to give you a general idea of what you can expect in the way of energy savings. You should realize that lower energy bills are just one benefit to installing a humidifier, and many folks elect to install them even though the energy savings will be marginal in their homes. It is true that in many installations a humidifier can save money for the homeowner. However, too often we have found that humidifier salesmen and dealers fail to tell their customers—perhaps because they do not realize themselves—that humidifier operation absorbs energy as the furnace air absorbs moisture. By making yourself aware of the fact that there is some heat loss accompanied with a humidifier's heat savings, you will avoid making a mistake that is all too common.

TYPES OF HUMIDIFIERS

There are several different ways to introduce moisture into household air and increase the relative humidity. Sometimes makeshift nonmechanical means are used to do this, such as boiling pans of water on the kitchen stove, placing open pans of water near heat registers, and draping wet towels over radiators. These may work all right as a temporary solution to low humidity problems, but it is much easier and much more efficient to install a humidifer in your house.

You can install either a portable humidifier as a separate unit in your house, or you can add a furnace humidifier to your furnace duct system. Either way, you get humidity added to the air on a regular, even basis, and you have a mechanical device that does all the work for you. The better humidifier installations have a humidistat to automatically maintain the relative humidity level set by the homeowner.

Portable Humidifiers

There is a wide variety of portable humidifiers available at most hardware stores, building supply stores and home supply centers. You can purchase units that are designed for single-room capacity only, or for entire house capacity. Generally, the larger the capacity, the more the humidifier will cost. Also, you can expect to pay more for a larger number of convenience features and for fancier cabinets.

One type of portable humidifier design is shown in Fig. 12-6. The spongy material on a rotating wheel picks up moisture from the water reservoir at the bottom of the humidifier. A fan pulls air from the living area and forces it through this filter pad. The air evaporates some of the moisture from the pad, and moist air is sent back into the living area.

Portable humidfiers plug into any 120 volt wall outlet. On most models you must fill the unit with water manually on a regular schedule according to the manufacturer's instructions.

TIGHTLY CONSTRUCTED HOUSE	FUEL SAVINGS: 3% × # OF DEGREES WHIICH THERMOSTAT IS LOWERED ↓ ↑ NET FUEL SAVINGS FUEL COST: 4000 BTU × ROOMS IN HOUSE	½ GALLON WATER EVAPORATED PER ROOM PER DAY
AVERAGE HOUSE	FUEL SAVINGS: 3% × # OF DEGREES WHICH THERMOSTAT IS LOWERED ↓ ↑ NET FUEL SAVINGS FUEL COST: 8000 BTU × # ROOMS IN HOUSE	1 GALLON WATER EVAPORATED PER ROOM PER DAY
POORLY SEALED HOUSE	FUEL SAVINGS: 3% × # OF DEGREES WHICH THERMOSTAT IS LOWERED ↓ ↑ NET FUEL LOSS FUEL COST: 16,000 BTU × # OF ROOMS IN HOUSE	2 GALLONS WATER EVAPORATED PER ROOM PER DAY. (EACH GALLON WATER EVAPORATED ABSORBS 8,000 BTU OG HEAT)

Fig. 12-5. The arrows pointing to the right in this diagram represent fuel savings from humidifier installation. Fuel savings from humidifier installation are about 3 percent of the fuel bill for each degree the thermostat is lowered. The arrows pointing to the left are energy costs from humidifier installation—that is, offsets against the energy savings. These energy costs are incurred because energy is expended when the water is evaporated into water vapor. As you will note, the amount of fuel expended varies greatly depending on how tightly the house is built. A tightly constructed house requires ½ gallon of evaporated water per day for each room. At that rate of water evaporation, not very much energy is absorbed in evaporating water, and there is a large net fuel savings. In the average house, where 1 gallon or water per day per room is required, there usually is still a net fuel savings from humidifier installation. But in a poorly sealed and insulated home, two gallons of water per room per day are needed to maintain the relative humidity at high levels. When this much water is evaporated, more energy is spent evaporating water than is saved by lowering the thermostat.

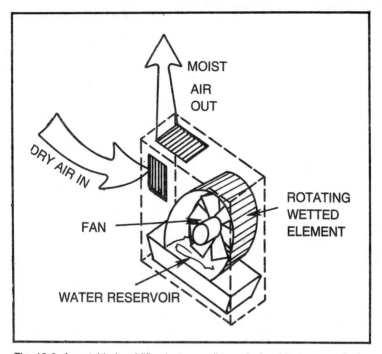

Fig. 12-6. A portable humidifier (cutaway diagram). A cabinet surrounds the humidifier unit. A rotating wheel covered with a sponge-like material absorbs water from the water reservoir. The fan draws air into the front of the humidifier and sends the moist air out the top. Generally, the water reservoirs on portable humidifiers must be filled manually.

Some come equipped with a hose that can be attached to a household faucet for filling. Some have humidistats to shut the unit off automatically when the desired relative humidity level is reached. A humidifier without a humidistat will operate continuously.

Furnace Humidifiers

In many respects, furnace humidifiers are similar to portable humidifiers. The largest difference is that furnace humidfiers are much more permanent installations, since they are installed in the duct system of the furnace. The advantage of having the humidfier installed in the duct system is that here the air is the hottest and has its largest moisture-holding capacity. Another advantage is that the duct system delivers the humidfier's moisture-treated air evenly to all rooms of the house, and not primarily to one or two rooms.

Furnace humidifiers do not have their own fans, since the furnace fan will circulate the humidfied air. They do not have a cabinet. The lack of these two items reduces their cost a bit. Furnace humidifiers also do not have to be filled manually, because they are normally attached to a water line by means of a hose and a saddle valve that taps the water line. They may be attached to a 120 volt circuit or a low voltage 24 volt circuit.

There are basically two types of furnace humidifers—evaporator types and spray types. The evaporator type furnace humidfier works in much the same way already illustrated for a portable humidfier. The spray type forces the water into the air stream in fine mist. Either way, the purpose of these humidfiers is the same—to introduce water vapor into the air stream of the furnace duct system.

Most furnace humidfiers can be installed by a do-it-yourselfer. Be sure the manufacturer's installation instruc-

Fig. 12-7. One design of evaporator humidifiers. This humidifier has rotating bristles that pick up water from the water basin and carry it up into the furnace air stream, which is over the top of the bristles when this unit is installed.

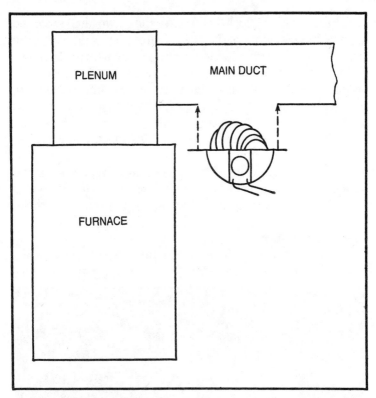

Fig. 12-8. Evaporator humidifiers are installed in the plenum chamber or in the main supply duct as near as possible to the plenum. A hole large enough for the humidifier element to extend through is cut in the furnace duct, and the humidifier attaches to the underneath side of the duct.

tions are clear and easy to understand. You can purchase a humidifier from a hardware store, a building supply store, or a furnace supply house. Later in this chapter we discuss humidifier installation.

Evaporator Types

Several different humidfier designs can be classified as evaporator-type humidifiers. Essentially what these humidifiers have in common is that they bring water into the furnace duct air stream with some type of wetted element. The hot furnace air passing over this element evaporates the water.

Figure 12-8 shows one common type of evaporator humidifier. This humidifier installs on the underneath side of the mainduct (Fig. 12-8). The bristles on the humidifier pick up water from the water basin. A 24 V electric motor rotates the bristles, and the wet bristles carry the moisture from the basin up into the furnace duct air stream, where the furnace air evaporates the moisture. Other types of humidifiers are available that use variations on this basic design—a rotating wetted element carrying water from a basin into the air stream.

Not all evaporator humidifiers use an electric motor. The plate-type humidifier shown in Fig. 12-9 merely has a pan filled with water installed in the duct system. Extending above the water pan are plates that absorb the water out of the pan. Air flowing through the plates evaporates the water. Plate-type humidfiers are installed in the plenum chamber where the air flow is the largest and the hottest.

Evaporator humidifiers have a reservoir filled with enough water to wet the humidifier element. A water line is attached to the reservoir through a valve to keep the reservoir filled. The water level is maintained with a float valve that opens and closes with the reservoir level to open or close the water line as needed.

Evaporator type humidifiers are among the least expensive humidfiers, but they are also among the least efficient. The plate-type humidifier in particular, without any motor to

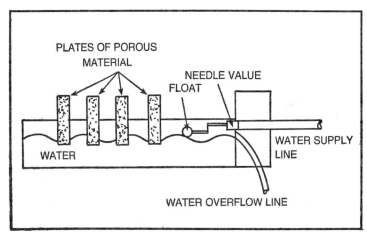

Fig. 12-9. One of the simplest types of humidifiers that installs in the furnace duct system is the plate-type humidifier. Plates made of a water-absorbing material are placed in a pan of water. The plates absorb the water and raise it up into the furnace air stream. The best location for a humidifier of this type is extended into the plenum chamber. This humidifier is also called a pan-type humidifier.

rotate the elements and increase the amount of water that gets into the furnace air stream, is quite inefficient. However, for small houses and light humidification applications, they are usually sufficient. But on the whole, for many homes evaporator humidifiers simply do not have the capacity to deliver enough moisture into the air. This is because they do not force the moisture into the air, as do the spray-type humidifiers, and they do not force the air to pass through the water-laden elements, as do belt type humidifiers.

One advantage to the evaporator type humidifiers is that mineral deposits and water impurities are left behind and do not find their way into the household air. With a spray-type humidifier, on the other hand, mineral deposits are constantly being sprayed out into the furnace air with the water, and these mineral deposits may appear in the house as fine white dust particles. Of course, since the mineral deposits are left behind in the evaporator humidifiers, this means scaly substances will build up on the humidifier parts. These mineral deposits must be removed periodically, as discussed later in the chapter in the section on humidifier maintenance. The presence of mineral deposits will create some inconveniences for humidifier operation in hard water areas.

Another Evaporator Type

Some of the operating inefficiency problems frequently accompanying the evaporator humidifiers already discussed are taken care of with the type of humidifier shown in Fig. 12-10. This type of humidifier is classified in several different ways, and some people place it and a family of similar humidifiers types into their own classification of furnace humidifiers. It may be called a bypass-type humidifier, a forced air humidifier, or a semi-forced moisture humidifier. What you call it really makes little difference. We have elected to place it in the category with evaporator type humidifiers because this humidifier still relies on the natural evaporation capacity of the furnace air to pick up the moisture.

There are significant differences in the way this type of humidifier works from those discussed earlier. This humidifier and those similar to it use an element made of sponge-like material to hold water. Then, part of the furnace air is diverted from the duct system and drawn directly through the wetted

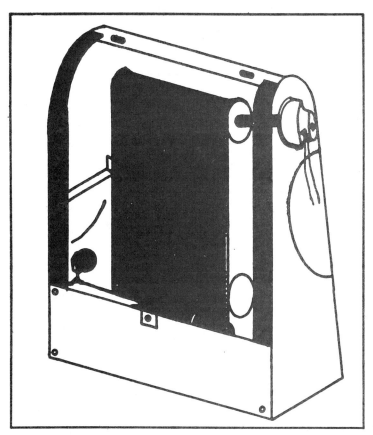

Fig. 12-10. This type of humidifier has a sponge-like belt held in place with two rollers. A motor turns the belt to saturate it with water in the reservoir at the bottom of the humidifier. A portion of the furnace air is diverted from the duct system and forced to pass through the wet element.

element. Sometimes a fan is used to divert the air and force it through the wet element, but the natural differences in air pressure within the duct system can also perform this task. As the air passes through the wet element, it picks up moisture by evaporating the water. Unlike the other evaporator types, the air is forced to pass through the wet element. The air does not simply pass around it. Most of these humidifiers have a water reservoir at the bottom of the humidifier housing where the element can pick up water. Some, however, have a system that wets the element by dropping water on it from above.

This type of humidifier is quite a bit more efficient than the simpler evaporator types. For this reason, it is absolutely necessary to install a humidistat when installing this humidifier. Otherwise, the humidifier will continue running even when the humidity is high, and the relative humidity will soon become too high, causing all sorts of problems.

Figure 12-11 shows how the belt-type humidifier is installed in the furnace. The humidifier unit is attached to the plenum chamber. A hole is cut in the return air duct. A flexible hose connects the return air duct with the humidifier and creates a bypass duct through which air will flow into the humidifier's element. Since the air pressure in the plenum is higher than that in the return air duct, some air in the plenum will be forced through the humidifier into the bypass. This air is forced through the saturated element and moist air goes up the bypass back into the return air.

Sprayer Types

The sprayer type humidifiers go under a number of different names, including atomization types and forced moisture types. These names, like the label "sprayer type," came about because of the way these humidifiers work. They use a pump to force the water into a fine mist (or atomized particles) that is quickly absorbed by the furnace air. Some models use a revolving disc to produce the water droplets that are absorbed.

The sprayer humidifiers are the most efficient humidifiers, and they are available in models that are an adequate size for virtually all humidifiers loads. By spraying the water directly into the furnace air stream, much larger amounts of water will be absorbed than when the air must evaporate the water from an element surface. Of course, there are two sides to that coin. While much higher levels of relative humidity can be obtained with a spray type humidifier, it is also easy for such a humidifier to introduce too much water into the air. A spray-type humidifier, therefore must be installed with a humidistat. Without one, the humidifier will just continue spraying more and more water into the duct system even though the air has reached the saturation point. This can cause problems outlined earlier that accompany too high relative humidity levels, and it can cause the duct system to begin rusting.

Mineral deposits may also cause some problems with these humidifiers in areas where the water is very "hard."

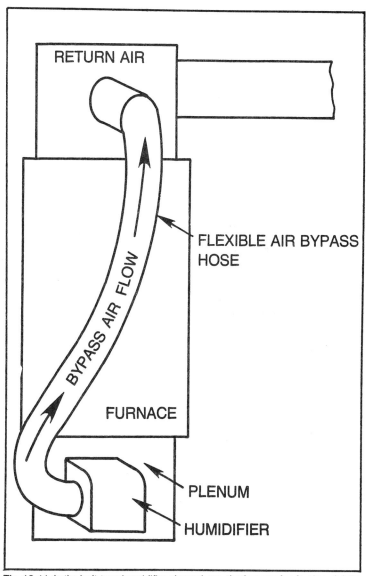

Fig. 12-11. In the belt-type humidifier shown here, the furnace fan forcing air into the plenum chamber combined with the low pressure in the return air duct, sends part of the furnace air through the humidifier element and into the bypass duct. Some similar systems use a separate humidifier fan to divert a portion of the furnace air and force it through the humidifier element. Because of the forced air flow, a humidistat should be installed with this type of humidifier.

With evaporator humidifiers, the mineral deposits are left behind on the humidifier surfaces. With a spray humidifier, on the other hand, those mineral deposits are sprayed directly into the air with the water mist. These deposits will form fine white particles that may appear in the living area as white dust.

SELECTION AND SIZING

As with almost everything else in a heating system, selecting a humidifier involves a number of compromises, and it's nearly impossible to say there is one "right" humidifier for any installation. One of the first things you'll have to do is determine what the capacity of the humidifier needs to be, because that may have quite a bit to do with what type is preferable. As we've already noted, some types of humidifiers are simply not designed to deliver large amounts of moisture into the air. Sizing the humidifier is discussed a bit later.

Once you have an idea of the size humidifier you want, there are several more considerations. Cost will be an important factor for most people, but it naturally should not be the sole factor considered. Don't be afraid to pay a bit more money for a humidifier that will have a capacity large enough for your home. Other factors are ease of installation and the amount of space available around the plenum chamber, return air duct or main duct for the humidifier apparatus.

Don't forget to consider ease of maintenance in selecting a humidifier. Mineral deposits that can affect the humidifier's operation will certainly build up, and they must be cleaned out periodically. Be sure there will be sufficient room for the humidifier to be removed and serviced as needed.

One of the simplest ways to size a humidifier is to determine the square footage or cubic footage of your home and match that with the capacity rating of the humidifiers. Most humidifiers are rated as capable of humidifying homes within a given range of square feet.

The problem with this method is that while it may be simple, it can also be overly simple. The square feet ratings can be assumed to be for homes of average construction, unless the rating says differently. If you have a tightly constructed home, a humidifier rated for the number of square feet in your home would be oversized by about double. Likewise, if you have a loosely constructed home, the humidifier would be undersized by about 50 percent. Sometimes more exact ratings are given with the humidifiers, such

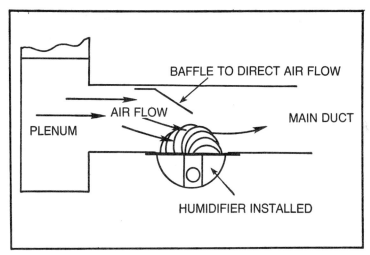

Fig. 12-12. A baffle should be made from sheet metal and attached to the top of the furnace duct to direct air flow over an evaporator humidifier. In some humidifier installation kits, this baffle is provided. If there is no baffle provided with your humidifier, fashion the baffle from the sheet metal you cut out of the duct to install the humidifier.

as capacity ratings for tight construction, average construction or loose construction. This helps you size the unit a little better for the home you have.

It's a good idea to buy a humidifier rated with extra capacity to humidify an area larger than your home. Many times the capacity ratings are actually high because of laboratory test conditions, or because of the assumed outdoor weather climate. Particularly if you live in a northern climate, you should purchase a humidifier with extra capacity because most of the ratings are based on outside climates in the southern half of the United States.

INSTALLATION

For portable humidifiers, no installation is involved. You simply bring the unit home, fill it with water and plug it in. That's all! Furnace humidifiers, on the other hand, must be connected to the furnace duct system. Actual installation instructions will vary according to the exact humidifier you purchase, but in this section we will try to give you a general idea of what to expect when installing a humidifier.

In virtually all humidifier installations there are three or four steps you will go through, and the steps are similar no matter what type of humidifier you have. You will have to cut a hole in the duct system large enough to attach the humidifier. You will tap a water line and connect water to the humidifier. You will wire the humidifier motor and fan (if the humidifier has a separate fan) into the furnace fan circuit. Finally, you will probably install a humidistat to control the humidifier motor.

To attach the humidifier to the duct system, you'll need a pair of tin snips and an electric drill. Cut the hole in the duct system just large enough for the humidifier to go inside the duct, but do not cut it so large that air leaks will be created. Attach the humidifier to the duct or plenum with sheet metal screws.

Some types of humidifiers are designed to be installed in the plenum chamber. The wetted element belt-type humidifier shown earlier in the chapter must be installed in the plenum chamber and an air bypass duct must be attached to the return air system for proper operation. The plate-type humidifer is usually installed in the plenum chamber where the air is warmest and can absorb the most moisture. The pan extending into the plenum must be level to keep the water from spilling out into the plenum and rusting the metal.

Other types of humidifiers, such as the rotating brush evaporator type already shown, are designed to be installed in the main supply duct near the plenum. These units attach to the underneath side of the main duct, and it is usually a good idea to try to install them as close to the plenum as possible, where the air is warmest. If the furnace has two or more supply ducts, install the humidifier in the largest one, if possible. You won't have to worry about uneven humidification through the house because the furnace system will distribute the humidified air evenly.

When you are installing a humidifier you must make sure water and power supply lines are reasonably close to the proposed humidifier location. Be certain the proposed humidifier location is such that if water overflows it will not damage the furnace, the burners, or create any electrical shock hazards. Remember, the float control systems and overflows on humidifiers are not infallible.

Be sure proper air will flow over the humidifier. Many humidifier manufacuturers provide a baffle with their humidifier installation kit to direct the air flow into the humidifier's elements. See Fig. 12-12. This is especially true with humidifiers that are installed in ducts. If your humidifier will be installed in a duct and does not have a baffle, you can make one easily out of sheet metal. When installing the unit, make sure gaskets and seals are tight.

The water supply line to the humidifier is usually ¼-inch copper, plastic or rubber tubing. First, however, you must have a water source. Water is supplied by tapping a nearby cold water line using a saddle valve similar to the one shown in Fig. 12-13. Sometimes these valves are supplied with the

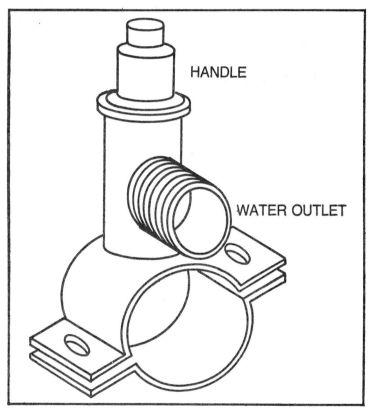

Fig. 12-13. A saddle valve is used to tap a water line and form a water supply for the humidifier. Specific directions for installing the saddle valve you use should come with the valve. If the valve is self-piercing, you merely tighten down on the screw-in handle to pierce the water line. You may have to drill a hole in the water line with some types of saddle valves. In any event, the clamps grasp the water line, and a tube from the valve goes into the water line. A gasket keeps the valve from leaking. You can attach the water line going to the humidifier at the outlet at the top of the valve.

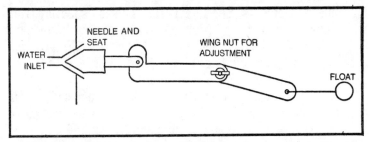

Fig. 12-14. When the water inlet is controlled by a needle valve attached to a float system, the arrangement will look something like this. Notice that when the float rises with the water level the float arm will push the needle into the seat to close the water inlet. When the float and water level lower, the valve opens. You can adjust the float level, and hence the reservoir level, by adjusting the float arm up and down. If a wing nut adjustment is not provided for this purpose, you can adjust the float by bending the float arm gently.

humidifier. If the valve supplied is the self-piercing type, all you usually have to do is attach the valve to the water line and tighten down on the valve handle to pierce the line. With some valves you may have to drill a hole in the water line first before the valve is installed. Complete instructions on how to install the saddle valve should be included with the saddle valve. Be sure the saddle valve is tight on the water line and the gaskets are tightly installed.

Once you have tapped the cold water line, you can connect the saddle valve to the humidifier water inlet with hose or tubing. Usually this inlet will be controlled with a float and needle valve or with an electrical solenoid. These devices control the water flowing into the humidifier. If your humidifier has a float control system, the water reservoir probably has a marking at the proper water level. You can adjust the float to achieve that water level by changing the arm adjustment, as shown in Fig. 12-14, or by simply bending the float's arm.

The humidifier reservoir has an overflow drain to keep the water level from getting too high. Any type of hose or tubing may be connected to this overflow connection. On some units a garden hose type fitting is provided. The hose should be run to any suitable drain, such as a sump pump or a sink.

WIRING THE HUMIDIFIER AND HUMIDISTAT

The final step to humidifier installation is connecting the wiring. Humidifiers will run off 120 volt and 24 volt circuits.

Usually it doesn't really matter what the humidifier voltage is, because the humidifiers are connected basically the same way. The humidifier is wired into the furnace fan circuit so that the humidifier can run only when the furnace fan is on. If you are installing a 120 volt humidifier, the humidifier motor is wired directly into the furnace fan circuit (also 120 volts), as shown in Fig. 12-15. If your humidifier has a 24 volt motor, the voltage in the fan circuit must be stepped down with a transformer, which normally is included as part of the humidifier assembly. In these installations, you simply connect the transformer leads to the furnace fan circuit. In some humidifiers, you may also have to connect the low-voltage transformer leads to the humidifier's motor. Figure 12-16 shows a 24 volt humidifier circuit.

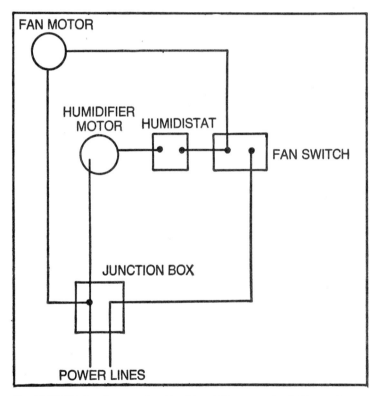

Fig. 12-15. This is a diagram of a 120 volt humidifier motor circuit. No transformer is used. Notice that when the fan switch closes, the humidifier circuit is also given power. Once the fan circuit closes, if the humidistat is calling for moisture and is also closed, the humidifier motor will start.

The humidifier is connected in the furnace fan circuit because you wouldn't want the humidifier to run unless there is hot air blowing through the duct system. At the very least, in evaporator humidifiers you'd be wasting the energy needed to run the humidifier, because no appreciable amount of moisture can be picked up when the furnace fan is not blowing. At the worst, a spray-type humidifier would spray its water mist into the duct system, and the water would collect and begin to corrode the ducts. Therefore, the humidifier is wired into the furnace fan circuit. Even when the humidistat calls for moisture, the humidifier will not operate unless the furnace fan is also turned on.

In most humidifier installations, the humidifier is wired in parallel with the furnace fan motor. In other words, one wire from the humidifier motor or transformer is connected to the fan side of the furnace fan control. The second wire is connected to the common wire that attaches to the other terminal on the furnace fan. See Fig. 12-17. As you will notice from looking at this diagram, you can connect the humidifier by attaching the two humidifier wires to the two terminals on the fan motor—if terminals are present. You can also trace the fan wire to the furnace fan control and attach one humidifier wire to the fan side of that switch. The furnace fan control is a temperature activated switch usually found in the heat exchanger. The second humidifier wire can be attached by following the common wire from the other fan motor terminal and attaching the second humidifier wire at the first terminal the common wire reaches.

The humidistat is wired in series with the humidifier motor. Without a humidistat, a humidifier will run whenever the furnace fan turns on. This is not necessarily bad. Some humidifiers, especially the evaporator types, will work reasonably well without humidistats. When there is enough humidity in the air, the air becomes saturated enough that it will not evaporate any more water. However, in any type of evaporator humidifier in which the air is forced through a water-laden element, a humidistat should be installed. Thus, the belt-type humidifier shown earlier in this chapter should have a humidistat because a portion of the furnace air is forced through it.

A humidistat must be installed in any type of forced

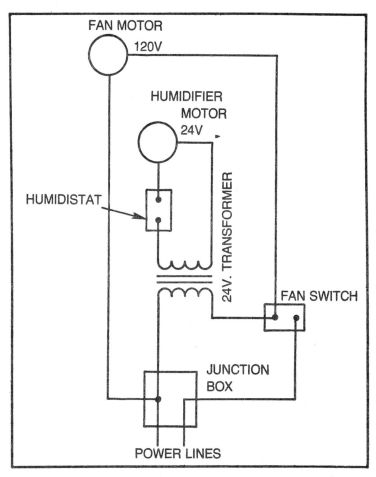

Fig. 12-16. When a 24 volt humidifier is used, a transformer must step down the 120 volt fan circuit to 24 volts. Normally this transformer is included as part of the humidifier. Notice that the humidistat merely forms a second switch in the humidifier circuit. The fan switch is the first switch. Unless the fan switch is closed, no power reaches the humidifier. Even if the humidistat is closed and calling for moisture, the humidifier will not run.

moisture or spray humidifier. Humidistats are necessities because in these installations the air will become too laden with water without them. If the air becomes saturated, these humidifiers will continue forcing water out into the airstream and causing water problems in the duct system, not to mention the problems saturated air casues in the house.

As noted earlier, the humidistat will not turn the humidifier on by itself. You can see from the wiring diagrams that the humidistat is merely a second switch in the humidifier circuit. First, the fan switch must be on before the humidstat can turn on the humidifier. And if the fan turns off, the humidifier will also turn off, even though the humidistat may sense the relative humidity of the air inside the house is too low.

The humidistat is a device that senses the amount of moisture in the air, and, according to the homeowner's setting opens and closes the switch to operate the humidifier motor. The humidistat can be either 24 volts or 120 volts. It will have the same voltage as the humidifier circuit. Humidistats may be located in the living area, or they may be installed in the return air duct or in the plenum chamber. Do not install the humidistat in the plenum chamber or return air duct if the humidifier is installed there. The humidity would be too high to give an accurate reading.

HUMIDIFIER MAINTENANCE

Evaporator type humidifiers usually require more maintenance than do sprayer types, simply because it is the mineral deposits left behind in the evaporator humidifiers that cause most maintenance requirements.

At least once a year before the heating season, you should remove the humidifier from the duct system and give it a thorough scraping and cleaning. If your water is especially hard, you may have to clean your humidifier at least once midway through the heating season.

Two of the most troublesome places for these mineral deposits to collect are on the float and the water inlet valve. Lime and other minerals caked onto the reservoir float can make the float so heavy that it will no longer float on the water. Naturally, this opens the valve and water pours into the reservoir unstopped.

Deposits on the needle and seat can clog the valve and make the valve stick open or closed. To clean the humidifier, turn off the water to the unit and drain the unit by opening the drain valve at the bottom of the reservoir. Remove the humidifier from the duct or plenum. Remove the humidifier element—belt, brush, plates, or whatever—and clean it as needed. Scrape the mineral deposits from the float, and the needle and seat on the inlet valve. Reinstall all the parts, and reconnect the humidifier. Be sure to get it level!

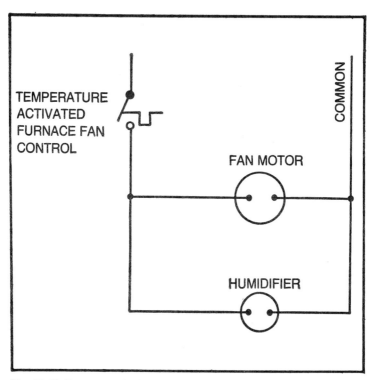

Fig. 12-17. To connect the humidifier, wire the humidifier in parallel with the furnace fan motor. This can be done by attaching one humidifier motor wire or transformer wire to the fan side of the furnace fan control and attaching the second motor or transformer wire to the common wire on the other fan terminal.

Humidifier Troubleshooting

Possible causes of high humidity are a malfunctioning humidistat, a humidistat located in dry place or set incorrectly and a water control valve stuck open. Possible causes of low humidity are mineral buildup on humidifier element, water turned off, humidistat set too low, humidistat stuck open, humidifier motor not working, water control valve stuck closed, loose electrical connections and a bad transformer. Possible causes of an overflowing humidifier are an improperly set float, a water control valve stuck open or the humidifier is not level.

Humidifier Summary

- Installing a humidifier will probably save you energy dollars if you live in a tight, modern house. It may save

you energy dollars if you live in a house of average tightness. It will probably cost you energy dollars if you live in a loosely constructed house.
- To save the most energy with a humidifier installation, the house should be tightly constructed so that little infiltration air enters the inside air. Good insulation, caulking, weatherstripping, storm door, windows and vapor barriers help retain humidity in the inside air.
- Forced moisture (spray-type) humidifiers are the most efficient humidifiers, followed by forced air humidifiers and evaporator humidifiers.
- Increase the efficiency of an evaporator humidifier by constructing a baffle to direct the air flow inside the duct into the humidifier.
- At least once a year, clean the humidifier and scrape away mineral deposits to insure proper humidifier operation.

Homeowner's Energy Tax Credit & Other Conservation Incentives

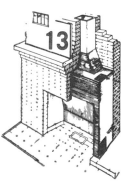

When Congress passed the National Energy Act in late 1978, it included a plan to allow homeowners to reduce their income tax bills when they install qualifying energy saving items. This plan is called the Residential Energy Tax Credit, and it presently includes energy-saving expenditures made from April 20, 1977 until 1986. Under the tax credit, a portion of the cost of qualifying energy saving items directly reduces your individual income tax bill on Form 1040.

The tax credit is available if you install items such as a set-back thermostat, an electronic ignition system to replace a gas pilot light on your furnace, or a more efficient furnace burner system. The credit is not available presently for expenditures such as wood heating systems, heat pumps or replacement furnaces. Items that qualify for the credit discussed in detail later in this chapter.

HOW THE CREDIT WORKS

Under the new residential energy credits, your income tax bill may be reduced by 15 to 30 percent of the cost of qualifying energy-saving items. Thus, when you take account of the income tax savings, the actual cost of the energy saving item is substantially reduced.

A 15 percent tax credit generally applies to the types of items we have covered in this book. This means that for every

$1 you spend on qualifying energy saving items, you can reduce your income tax bill by 15 cents. That's a direct reduction—or credit—of your income tax bill of 15 cents per dollar spent, no matter what your tax bracket is. The tax *credit* is different than a *deduction*, so you get the credit even though you do not itemize your deductions.

There is a second energy tax credit, also. This is a higher credit of 30 percent of the cost of "renewable energy source" items, such as solar energy, geothermal energy or wind energy items. Since we have not discussed these types of devices in this book, we will not dwell on the 30 percent energy tax credit here. But if you are considering installing any of these "renewable energy source" items, you should check with your tax preparer to get the details on this credit. It can be a substantial savings.

Two thousand dollars is the maximum amount of expenditures eligible for the 15 percent credit. Thus, your maximum tax credit is $300. This limit applies for the time you own your home. If you spent $500 and took a $75 tax credit this year, you may claim a maximum credit of $225 during the following years. You might think of the tax credit limit as a sort of "bank" that gives you a beginning balance of $300 in tax credits available for $2000 in expenditures. Year by year, as you take an energy tax credit on your income tax, you reduce your balance in the "bank" by the amount of credit you have declared.

You can declare the entire $300 tax credit limit in one year, if you make $2000 in qualifying expenditures, or you can stretch the $300 "balance" out over several years. If you move into a new home, you start fresh with a new $300 limit.

To figure the amount of your credit, you simply take the cost of the qualifying item, add in the cost of labor for installation (if you did not install the item yourself), and multiply the total cost by 15 percent. This is the amount of your tax credit.

For example, suppose you install this year an automatic set-back thermostat. Assume the cost of the thermostat is $75, and the cost of installation is $35. Your tax credit would be:

$$\$75 + \$35 = \$110 \text{ total cost,}$$
$$\$110 \times .15 = \$16.50 \text{ tax credit.}$$

Thus, your tax credit on this item is $16.50, which is added to your tax credit for any other qualifying purchases for this year. If you had no other qualifying expenditures, your income tax bill would be reduced by the amount of the credit—$16.50.

SOME BASIC RULES

In order to qualify for the energy tax credit, there are a number of rules you must meet. First, the energy saving items must be installed on your principal residence. Thus, installations on vacation homes do not qualify. The items must be new. The purchase and installation of used items does not qualify. If your total tax credits in a year are less than $10, you will not be eligible to claim any credit at all for that year. Therefore, if you are considering making several small energy saving purchases, you might want to group several of them together to be purchased in the same year to put you above the $10 minimum.

The energy saving item must have an expected life of at least three years to qualify for the tax credit. You may claim the tax credit only in the year in which the qualifying item was installed, not when purchased.

The home on which the item is installed must have been completed—or at least "substantially completed" according to Congress—before April 20, 1977.

You should save all receipts, cancelled checks and records of your expenditures to justify your claim of the tax credit in case of an audit.

If you were eligible for a tax credit but did not declare it in prior years, you cannot aggregate those expenditures with the current year and take the credit on this year's tax return. However, you may be able to file an amended return for those earlier years. If you believe you were eligible in prior years for a tax credit that you did not declare, check with your tax preparer.

ITEMS THAT QUALIFY

The Residential Energy Tax Credit plan that passed Congress in 1978 lists a number of items that qualify for the 15 percent tax credit. It is likely that in the future—additional items will be added to this list, so some of the items that do not qualify for the credit as of this writing may qualify in time. Also, future IRS regulations may specify certain models of items that qualify for the credit. As of this writing, no such regulations had been issued.

If you are considering installing an energy conserving item, it would be a good idea to contact your tax preparer or your local IRS information office to determine if the item you are considering is covered under the tax credit. Qualifying and non-qualifying items may change rapidly, so you should check to be sure.

Under the tax credit plan, these items are listed as items that qualify for the 15 percent tax credit:

- Meters that display the cost of energy usage.
- Automatic flue dampers. These are devices installed in the furnace flue that automatically close the flue when the furnace is off.
- Electrical or mechanical ignition systems that replace the pilot light in a gas heating system.
- Oil or gas furnace replacement burners that increase combustion efficiency and reduce the amount of fuel consumed.
- Furnace duct or water pipe insulation.

There are a number of other items that also qualify for the tax credit, such as insulation, storm windows, etc., but they are not connected with the subject matter of this book and will not be covered here.

These items do **not** qualify for the energy tax credit as of this writing:
- Replacement furnaces or boilers.
- Heat pumps.
- Wood burning stoves or furnaces.
- The cost of annual furnace tune-ups.

However, as we noted earlier, there are a number of regulations soon to be issued that may change the picture. Some of the items that do not now qualify may qualify by the time you install them. If you are considering expenditures on any of the above items, contact your tax preparer or your local IRS information office. You may be pleasantly surprised.

STATE AND LOCAL GOVERNMENT ENERGY INCENTIVES

As of this writing, the Energy Tax Credit is the most important federal incentive plan for energy conservation available to homeowners. However, many states have taken a keen interest in residential energy conservation and have enacted plans of their own to encourage it. The plans vary widely from state to state, but most include tax breaks of some type for expenditures on qualifying items. Some states provide reductions of the state income taxes. In others, no sales taxes are applied to purchases of qualifying items. In still others, property tax abatements or reductions are available. The best way to find out what incentives are available in your area is to contact your state's energy office.

Glossary

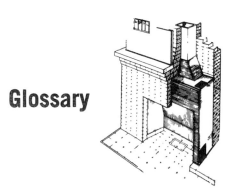

air filter—A central warm-air furnace will have one or more air filters to remove the dust, dirt and lint from the air flow. There are three types of air filters: disposable, washable and electronic. Disposable filters are made to be used once and thrown away when dirty. Washable filters are made of a foam-like element that may be washed out when dirty. Electronic air filters—the most effective type—use a large electrical charge to attract microscopic dust particles from the air as it passes between two highly charged electrical plates.

amperes—A measure of electrical current. Measured with an ammeter. Also called amps.

anticipator—A small resistor that warms the thermostat to make the thermostat turn off the furnace burners about 1° early. Thus, when the fan blows the residual heat out of the heat chamber, the room temperature will climb the remaining 1° to the thermostat setting, but will not "overshoot" the thermostat setting.

balance point—The temperature at which a heat pump's refrigeration system is no longer able to supply the home's heat requirements and supplementary heat must be added.

barrel—The portion of an oil burner that extends into the combustion chamber. Houses electrodes and nozzle assembly.

baseboard heater—A compact heating unit that is installed along the baseboard in a room. The baseboard heater, usually an electric resistance heater, provides heat for a single room.

belt driven fan—A furnace fan in which the electric motor driving the fan is connected to the fan blade shaft through pulleys and a belt.

bimetal switch—A heat-sensitive control made by fusing two metals together. The resulting combination will bend with temperature changes. Bimetal switches are used often in thermostats and in limit controls.

boiler—The part of the hydronic heating system containing the combustion system. The system's water is heated in the boiler.

boot—The duct piece that sends the furnace air from the runs into the living area. The boot is connected to the room register.

ceiling cable—Electrical resistance heating elements installed in ceiling panels.

central heating system—One furnace supplies heat through a duct system to every room in a building.

circulating pump—A pump that circulates water through a forced-water hydronic heating system. Not all hydronic systems have circulating pumps. Some depend on the natural tendency of warm water to rise to circulate water through the system.

circulating stove—A stove that has the firebox contained within an outer jacket. Air can circulate between the outer walls of the firebox and the jacket. In circulating through this area the air is warmed before returning to the living area.

closed circuit—A circuit through which electricity can flow.

coefficient of performance(COP)—A figure used to rate the efficiency of heat pumps on the heating cycle. COP = BTU per hour divided by watts per hour.

coil case—The part of the furnace's body that houses the evaporator.

combination furnace—A furnace that burns two fuels, generally wood and a supplementary fuel.

combustion chamber—The part of the furnace where the actual combustion process that produces heat takes place. The combustion chamber is sealed from the air that flows into the house.

compressor—A mechanical component of a refrigeration system that moves refrigerant through the system to transfer heat.

condensation—When water vapor changes from a gaseous state to a liquid state in the form of water droplets. Condensation occurs when warm moisture laden air comes in contact with a cold surface. At the colder temperatures, the air cannot hold all the water vapor, and the excess vapor condenses to form water.

condenser—A refrigeration system coil where refrigerant is compressed and changes from a gaseous state to a liquid, giving off heat as it does so. The condenser is the warmer coil in the system.

connector—The stovepipe that joins the stove to the chimney.

continuity tester—A device that determines whether electricity is flowing between two points. An ohmmeter is often used for this purpose.

cooling coil—Also known as the evaporator. The part of an air conditioning system or a heat pump where heat is removed from the surrounding air.

cord—A common measure of firewood. One cord is a stack of firewood 4' × 4' × 8'. One cord of seasoned hardwood has about the same heat potential as one ton of coal.

creosote—Tar-like substance produced by burning wood, especially unseasoned wood and softwood. Creosote collects in the chimney and is highly flammable.

damper—A movable plate that controls the draft and burning rate in wood or coal-burning devices.

dehumidifier—A device that removes moisture from the air.

direct drive fan—A furnace fan where the electric motor is connected directly to the fan blade shaft. That is, the fan blades are installed on the motor shaft.

downflow furnace—A furnace in which the air flows from top to bottom.

dry bulb temperature—The temperature as measured with an ordinary thermometer. Does not take into account the effects of humidity and evaporation on the apparent temperature. Compare to wet-bulb temperature.

EER—See Energy Efficient Ratio.

electrodes—The electrical devices at the nozzle of a gun-type oil burner that arc to light the oil.

electric resistance hat—Heat produced by electric furnaces and room heaters. Heat is produced by sending electrical current

electric resistance hat—hardwood

through metal elements that have a certain amount of resistance to electricity flow. When current is sent through the elements, this resistance produces heat.

energy efficiency ratio—A figure that is used to compare the efficiencies of air conditioners. EER = BTU per hour divided by watts per hour.

evaporator—A refrigeration system coil where refrigerant expands and changes from a liquid to a gaseous state and absorbs heat. The evaporator is the cooler coil in the system.

expansion tank—A tank located near the boiler on a hydronic heating system. The tank is half filled with air and half filled with water. When the heated water in the system expands, more water fills the expansion tank and air in the tank compresses.

fan switch—The control that turns the furnace fan on and off. When the temperature in the heat chamber reaches the turn-on temperature, the switch closes and starts the furnace fan. After the furnace burners quit, the fan continues blowing until the temperature in the chamber drops to the turn-off temperature.

feedback—As applied to electronic troubleshooting, feedback occurs when the service technician has not properly disconnected the circuit to make a continuity check. Thus, because another circuit is connected and there is a path for electricity to flow through, a continuity reading results, even though the circuit he intends to check is open.

firebox—The portion of a combustible-fuel stove or furnace where the fuel burns. Usually refers to the combustion chamber of a wood or coal furnace or stove.

flue heat exchanger—A device that attaches in line with the flue of a stove or furnace to recapture heat escaping through the flue.

gas valve—The device on a gas burner that controls the flow of gas into the burners. The gas valve is activated by the thermostat to open and close the flow of gas.

gun type oil furnace—An oil furnace in which the fuel burns as it is sprayed between igniters.

green wood—Also known as unseasoned wood. Undried wood cut from a live tree.

hardwood—Wood varieties that possess good burning qualities. Includes woods such as oak, maple, and hickory, among others.

heat-circulating fireplace—Also called a circulator fireplace. A fireplace that has a double-wall steel lining in the fire chamber. Air circulates inside this lining, is warmed, and exits into the living area.

heating-only thermostat—A thermostat to be used only with a heating system. Cannot be used with central air conditioning.

heating-cooling furnace—A temperature control system that has a furnace to heat during the winter and an air conditioner to cool during the summer connected into the same duct system.

heat exchanger—A device that transfers heat from a warm area to a cool area. It often refers specifically to the area of the furnace where the air is warmed.

heat pump—A heating device that uses refrigeration principles to transfer heat from outdoor to indoor air.

horizontal furnace—A furnace in which the components are arranged side-by-side instead of stacked on top of one another.

humidifier—A device that automatically adds moisture to the air in a house. Portable cabinet-type humidifiers and furnace humidifiers that install in the duct system are available.

humidistat—A device that senses the amount of relative humidity in the air and automatically controls the humidifier to provide more moisture or less moisture as needed.

humidity—The amount of moisture in the air.

hydronic heating—A heating system that circulates water, rather than air, to transfer heat from the heating unit to the living area. A hydronic heating system includes a boiler, room registers and connecting water pipes.

igniters—see electrodes.

infinity—As used in electrical troubleshooting, the reading on the far left of an ohmmeter scale indicating "infinite" resistance and an open electrical circuit.

limit switch—A protective device in a furnace that shuts down the furnace when the heat chamber gets too hot. This device protects the furnace and house from damage if the burners begin heating and the fan does not come on for some reason to take the heat into the living area.

line voltage thermostat—A thermostat that is wired directly into the power line going to the heating unit. 120 volt and 240 volt thermostats are line voltage thermostats.

liquid petroleum (LP) gas—A heating gas that is produced by blending petroleum products manufactured from oil. It is stored under pressure in tanks.

low voltage thermostat—The type of thermostat most often used on central heating systems. The thermostat is wired into a low-voltage line (usually 24 volts) that is created by a transformer within the furnace that reduced the thermostat circuit voltage.

main duct—The duct that extends from the plenum chamber in a furnace duct system to supply the individual runs to the rooms. There may be more than one main duct in some duct systems. Also called main trunk line and supply duct.

manifold—The part of a gas burning combustion system that carries gas from the gas valve to the burners.

natural gas—A heating fuel found underground that is sent through pipelines to the consumers.

nozzle—The part of a gun type oil burner that sprays the oil into the combustion chamber under high pressure.

ohms—A measure of electrical resistance. Measured with an ohmmeter.

ohmmeter—A device that checks the amount of resistance between two points. Also may be used as a continuity tester to determine if electricity is flowing between two points.

open circuit—Describes a circuit through which no electricity can flow. Electricity can flow only through a closed path. In an open circuit, the path is broken and no electricity flows.

orifice—The device attached to the manifold of a gas burner that meters gas into the burner. By changing orifices, you can change the amount of gas flowing into the gas burner.

outdoor thermostat—A thermostat that senses the outdoor temperature and will not allow heating elements to turn on until the temperature is below a preset level. Used most often with heat pumps and electrical furnaces to limit the number of heating elements that will turn on at relatively mild temperatures.

plenum chamber—The first part of the duct system the air enters upon leaving the furnace.

pilot light—A continuously burning flame on a gas furnace that lights the burners once gas begins flowing to them.

pot-type oil furnace—An oil furnace in which a pool of fuel oil burns as it vaporizes.

pressure regulator—A device that can be adjusted to change the amount of gas pressure in a gas burning furnace or hydronic system.

radiant stove—A stove that achieves almost all its heat transfer by radiation instead of by convection air currents, as does a circulating stove. A radiant stove does not have an outer jacket through which air can circulate to transfer heat.

radiator—A room heat register of a hydronic heating system. Hot water or steam circulate through the radiator and transfer heat to the surrounding air.

refrigerant—Material that fills a refrigeration system and changes from a liquid to a gas and back to a liquid to transfer heat. When it changes from a liquid to a gas, refrigerant absorbs heat at the evaporator coil. When it changes from a gas to a liquid in the condenser, refrigerant releases heat.

relative humidity—The amount of moisture held by the air compared to the amount of moisture the air is capable of holding at that temperature. Thus, 50 percent relative humidity means the air is holding half its moisture-holding capacity at that temperature.

resistance heat—See electric resistance heat.

return air duct—The duct that removes circulated air from the living area and returns it to the furnace to be reheated or re-cooled.

reversing valve—A device on a heat pump that changes the direction the refrigerant flows through the system to change the heat pump from its heating to its cooling cycles.

runs—The individual ducts attached to the main duct of a furnace duct system. The runs take the furnace air from the main duct to the rooms.

saddle valve—A valve that attaches to a water line to tap the line and take off water from it.

seasonal performance factor (SPF)—A figure showing the heat pump's efficiency over an entire heating season. Relates total heat energy output to energy input.

seasoned wood—Also known as dry wood. Wood that has had most of the natural moisture removed.

self-contained heat pump—A heat pump where both coils are housed in one unit that is placed outside the house and connected to the house's duct system through the foundation.

set-back thermostat—A thermostat that can be set to turn down the house temperature at night or when the house is empty, and then bring the temperature back up before anyone wakes up or comes home. Some set-back thermostats can be set to turn down the house temperature for several periods a day, and for different periods on different days of the week.

sequencer—A 24 volt electrical relay control that controls the heating elements of an electric furnace.

short circuit—A circuit that does not follow the intended electrical path. A common example is a loose wire touching a case or housing to cause a short circuit.

short cycling—A furnace malfunction where the fan turns on and off in short intervals.

single-stage thermostat—A thermostat that brings all the furnace's heating capacity on at once, as opposed to a two-stage thermostat.

softwood—Wood varieties that possess less desirable burning qualities than hardwoods. Includes pine, fir, and cedar, among others.

space heater—Portable heater, often electric, that may be used to heat a small space.

SPF—see seasonal performance factor.

split system—A heat pump system where each coil and fan is housed in a separate indoor and outdoor unit.

staging—When a furnace is connected to a two-stage thermostat so that different portions of the heating system come on at different times.

stoker—A device on a wood or coal furnace that delivers fuel to the combustion chamber.

strip heaters—The electrical resistance heating elements of a heat pump system.

take-off—A duct connection that connects the run ducts to the main duct.

thermocouple—A safety device that senses heat from the pilot light. As long as the pilot light is burning, the thermocouple will allow the gas valve to open. But if the pilot light goes out, the thermocouple will not allow the gas valve to open.

thermostat—The heat-sensitive switch that turns the furnace and air conditioning system on and off.

thimble—A device to protect combustible walls and ceilings from heat when stovepipes or chimneys pass through structural parts of the house. The thimble itself is a sleeve that the stovepipe or chimney passes through.

timed thermostat—See set-back thermostat.

two-stage thermostat—A thermostat that brings about half the furnace's burners or heating elements on when the thermostat first calls for heat. If the temperature of the living area continues to drop, the second stage of the thermostat is activated and the second half of the heating capacity comes on. Two-stage thermostats are often used as energy-saving thermostats for electric furnaces and heat pumps where it is very inefficient to bring the entire heating capacity on at once.

upflow furnace—A furnace in which the air flows from bottom to top.

vapor—The gaseous state of any matter.

voltmeter—A device that tests the voltage, or electrical power, reaching two terminals.

volts—A measure of electrical power. The most common voltages used in heating system circuits are 240 volts, 120 and 24 volts.

wet bulb temperature—The temperature as measured with a thermometer that takes account of the fact that evaporation makes the temperature feel cooler. A wet-bulb thermometer has a moistened cloth wick wrapped around the bulb so that the evaporation of water from the bulb registers the apparent temperature.

Appendices

Appendix A

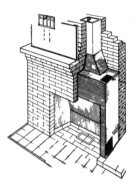

The following companies have provided pictures and illustrations that appear in this book. Their complete addresses appear here so you can contact them for further product information if you desire to.

Borg-Warner Corporation York Division,
Box 1592,
York, Pa. 17405.

Dolin Metal Products,
475 President St.
Brooklyn, NY 11215.

Edwards Engineering Corporation,
Pompton Plains, N.J. 07444.

Intermatic, Inc.,
Intermatic Plaza,
Spring Grove, Ill. 60081.

Kickapoo Stove Works, Ltd.,
Box 127-2N,
LaFarge, Wisconsin 54639
(608) 625-4430.

Longwood Furnace Company,
Gallatin, Mo. 64640

Monarch Ranges and Heater Division of the Malleable Iron Range Company Wood Stoves,
Beaver Dam, Wisconsin 53916;
(414) 887-8131.

Shenandoah Manufacturing Company,
P.O. Box 839
Harrisonburg, Va. 22801.

The Williamson Company,
3500 Madison Road,
Cincinnati, Ohio 45209.

Appendix B

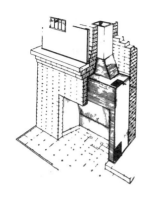

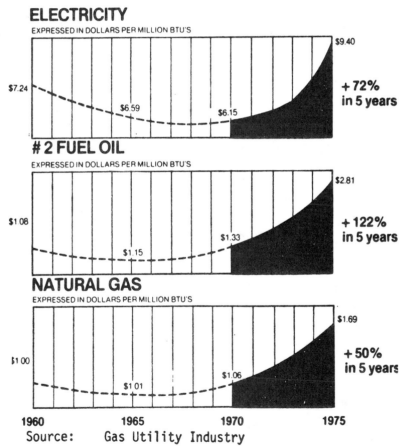

Fig. B-1. History of home energy costs.

	Supply Duct Size (Inches)		
Flow, cfm	Round	Rectangular	
100	6	8 × 6	6 × 8
125	6	8 × 6	6 × 8
150	7	8 × 6	6 × 8
175	8	8 × 6	6 × 8
200	8	8 × 8	
225	9	8 × 8	
250	9	8 × 8	
300	10	8 × 10	10 × 8
350	10	8 × 10	10 × 8
400	10	8 × 12	10 × 10
450	12	8 × 14	10 × 10
500	12	8 × 14	10 × 10
600	12	8 × 14	10 × 12
700	12	8 × 16	10 × 14
800	14	8 × 20	10 × 16
900	14	8 × 22	10 × 18
1000	16	8 × 24	10 × 20
1200	16	8 × 28	10 × 22
1400	18	12 × 22	10 × 26
1600	18	12 × 24	10 × 28
1800	20	12 × 26	10 × 32
2000	20	12 × 28	10 × 34
2500	22	12 × 32	10 × 40
3000	24	12 × 38	14 × 32
3500	26	12 × 42	14 × 38
4000	28	14 × 44	16 × 38

Fig. B-2. Required duct size as a function of air flow.

Wire Size	Max. Load in Amps	Max. Load in Amps.
	copper wire	aluminum wire
14 AWG	15	NA
12 AWG	20	15
10 AWG	30	25
8 AWG	40	30
6 AWG	55	40
4 AWG	70	55
3 AWG	80	65
2 AWG	95	75
1 AWG	110	85

Fig. B-3. Current-carrying capacity of copper and aluminum wire.

(1) Size AWG MCM	(2) Rubber Type R, Type RW, Type RU, Type RUW (14-2) / Type RH-RW / Thermoplastic Type T, Type TW	(3) Rubber Type RH / Type RH-RW / Type RHW	(4) Paper / Thermoplastic Asbestos Type TA / Var-Cam Type V / Asbestos Var-Cam Type AVB / MI Cable
14	15	15	25
12	20	20	30
10	30	30	40
8	40	45	50
6	55	65	70
4	70	85	90
3	80	100	105
2	95	115	120
1	110	130	140
0	125	150	155
00	145	175	185
000	165	200	210
0000	195	230	235
250	215	255	270
300	240	285	300
350	260	310	325
400	280	335	360
500	320	380	405
600	355	420	455
700	385	460	490
750	400	475	500
800	410	490	515
900	435	520	555
1,000	455	545	585
1,250	495	590	645
1,500	520	625	700
1,750	545	650	735

Fig. B-4. Current ratings of insulated wires.

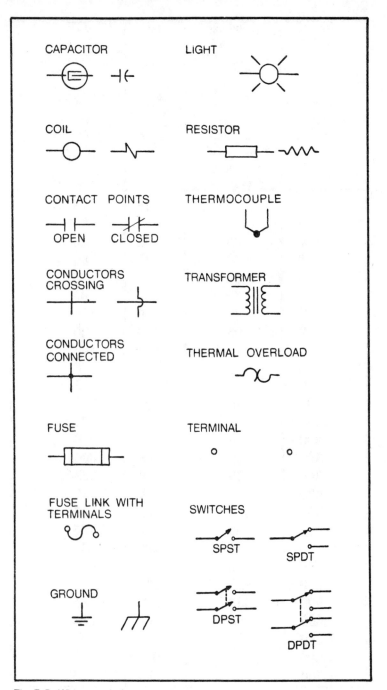

Fig. B-5. Wiring symbols.

OUTDOOR RELATIVE HUMIDITY	-20°	-10°	-5°	0°	+5°	+10°	+15°	+20°	+25°	+30°	+35°	+40°	+45°	+50°
100%	2%	3%	4%	6%	7%	9%	11%	14%	17%	21%	26%	31%	38%	46%
95%	2%	3%	4%	5%	7%	8%	10%	13%	16%	20%	24%	30%	36%	44%
90%	2%	2%	4%	5%	6%	8%	10%	12%	15%	19%	23%	28%	34%	41%
85%	2%	2%	4%	5%	6%	8%	9%	12%	15%	18%	22%	27%	32%	39%
80%	2%	2%	4%	5%	6%	7%	9%	11%	14%	17%	20%	25%	30%	37%
75%	2%	2%	3%	4%	5%	7%	8%	10%	13%	16%	19%	23%	28%	36%
70%	1%	2%	3%	4%	5%	6%	8%	10%	12%	15%	18%	22%	26%	32%
65%	1%	2%	3%	4%	5%	6%	7%	8%	11%	14%	17%	20%	25%	30%
60%	1%	2%	3%	4%	4%	5%	7%	8%	10%	13%	15%	19%	23%	28%
55%	1%	1%	2%	3%	4%	5%	6%	8%	9%	12%	14%	17%	21%	25%
50%	1%	1%	2%	3%	4%	4%	6%	7%	9%	10%	13%	16%	19%	23%
45%	1%	1%	2%	3%	3%	4%	5%	6%	8%	9%	12%	14%	17%	21%
40%	1%	1%	2%	2%	3%	4%	4%	6%	7%	8%	10%	12%	15%	18%
35%	1%	1%	2%	2%	3%	3%	4%	5%	6%	7%	9%	11%	13%	16%
30%	1%	1%	1%	2%	2%	3°	3%	4%	5%	6%	8%	9%	11%	14%
25%	1%	1%	1%	1%	2%	3%	3%	3%	4%	5%	6%	8%	10%	12%
20%	,%	1%	1%	1%	1%	2%	2%	3%	3%	4%	5%	6%	8%	10%
15%	÷%	1%	1%	1%	1%	1%	2%	2%	3%	3%	4%	5%	6%	7%
10%	÷%	÷%	1%	1%	1%	1%	1%	1%	2%	2%	3%	3%	4%	5%
5%	÷%	÷%	÷%	÷%	+%	+°	1%	1%	1%	1%	1%	1%	2%	2%
0%	0%	0%	0%	0%	0%	0%	0%	0%	0%	0%	0%	0%	0%	0%

OUTDOOR TEMPERATURE

Fig. B-6. Outdoor-indoor relative humidity conversion chart.

Index

Index

A

Air
  bleeder valve 60
  flow, check out 121
  flow through the furnace 32
  mixture, adjusting 143
Ammeters 23, 24
  checking malfunctioning
    electric motor 28
  clip-on 24
  types 25
Amperes 24
Amps 24

B

Balance point 254
Baseboard
  heat 53
  heater, installing 177
  heater troubleshooting 181
  heater tuneup 181
Bearing lubrication 94
Bimetal switch 45
Bladed fan 84
Boilers 56
  combustion process 57
  fuel-powered 31
British Thermal Units rating 31
BTU rating 31
Burners 34, 49
  cleaning 118
  gas 132

  servicing 118
  spread-type 134

C

Ceiling cable heaters 53
Central
  air conditioning 40
  heating system, attaching
    a stove to your 224
Circuits 21
  closed 21
  open 21
  short 21
Circulating
  pump 59
  stoves 214
Chamber 32
  combustion 32
  fire 231
  heat 32, 38, 49
  plenum 36, 38
Checking the circuit 20
Chimney 215-219
Cleaning solvents 92
Clip-on ammeters 24
Closed circuits 21
Coal 46
  furnaces 46-48, 156, 226
  stoves 206
Coil cases 34, 38
Combination furnaces 53, 162, 228
Combustion chamber 32

Condensing unit	40
Connectors	215-219
Continuity tester	16
making own	20
storing	20
Cooling coil	34
Cord	70
Credit	321
basic rules	323
how it works	321
items that qualify	323

D

Damper	71, 232
Direct drive squirrel cage fan	84
Disposable furnace filter	76
Downflow horizontal furnaces	38
Duct	
registers & their placement	190
system basics	183
system, insulating	196
tools	188
Ductwork	189

E

Electric	11
drill	11
furnaces	146
furnaces, operation	147
furnaces, servicing	147
heat	52
ignition	134
room heaters, installing	175
Electrical tape	11
Electricity	31, 46
Electrodes	51
Electronic furnace filter	78
Energy	
efficiency checklist	106
savings	189
Evaporator	34
Evaporator coil	40
Expansion tank	60

F

Fan	32, 38, 108
bladed	84
blades, cleaning	88
cleaning	81
control, adjusting	96
direct drive squirrel cage	84
removing	84
squirrel cage	82
types	82
Fan motor	84, 108
cleaning	88
disassembling	90
reassembling	96
removing	84
Fire chamber	231
Fireplace	206, 230
checklist	241
glass doors	237
heat output, increasing	236
parts	231
prefabricated	235
safety	234
Flue	232
Forced water systems	56
Fuel	
costs	65
costs, formula	66
costs, some examples	68
efficiency & heat pumps	72
filter, cleaning	116
-powered boiler	31
Furnace	11, 206
coal	46-48, 156, 226
combination	53, 162, 228
duct tape	11
electric	146
gas	163
LP gas	48
natural gas	438
oil	50, 110, 164
operation	46
tuneups	104
wood	46-48, 71, 226
Furnace filter	36, 74, 106
disposable	76
electronic	78
location	75
types	75
washable	77

G

Gas	31, 46
burners	132
burners, cleaning	138
burners operation	132
burners, servicing	145
burners, troubleshooting	145
burners tuneup	136
burners tune-up checklist	144
furnaces	163
pressure, adjusting	140
valve	48, 136
Grates, hollow-tube	238

H

Hardwoods	70
Hearth	232
Heat	32
chamber	32, 38, 49
exchanger	32

load	31
registers	58
Heat pumps	54
buying tips	262
efficiency	260
defrost cycle	251
increasing popularity	243
maintenance	263-265
metering devices	250
operation	246-250
reversing valve	251
types	260
valves	250
will it save money	244
Heated crawl space	198
Heater size, selecting	176
Heating	
-cooling funaces	40
-cooling furnaces, duct system	42
-cooling furnaces, thermostat	42
systems	31
systems, adding coal heaters	159
systems, adding wood heaters	159
system fuels	31
system fuels, electricity	31, 46
system fuels, gas	31, 46
system fuels, oil	31, 46
system fuels, wood	31, 46
systems, hot water	31, 54
systems, steam	31, 54
Hot water heating systems	31, 54
Humidifiers	291
& fuel savings	297
installation	311
maintenance	318
selection	310
sizing	310
wiring	314
types	300-310
Humidistat	296
Humidistat, wiring	314
Humidity basics	292
Hydronic	
heating basics	54
system, other controls	61
system, other valves	61

I

Igniters	128
Infinity	16
Installing a new furnace	165-171

K

Kilowatts	32

L

Lifetime bearings	94
Limit	45
control	130
switches	45, 112
Liquid petroleum	31, 46
Living area	36
Local government energy incentives	325
LP gas furnaces	48

M

Main supply duct	36
Masonry fireplace	71
Motor shaft	94

N

Natural	
gas	31, 46, 132
gas furnaces	48

O

Ohmmeters	16, 28
continuity tester	16
has own power	16
using	16
Oil	31, 46
burner tune-up checklist	127
delivery problems	130
Oil furnaces	50, 110, 164
electrical system	112
filter	114
fuel tank	114
gun type	51
operation	110
pot type	51
safety controls	111
tuneups	114
Open	18
Open circuits	21
Overheating	45

P

Pilot light	134, 140, 144
Plenum chamber	36, 38
Pliers	11
Pop-rivet gun	11
Protectors	148
overload	148
wiring	148
Pump	120
filter	120
pressure, check out	121

R

Radiators	58
Registers, adjusting	195

Return air duct	36, 38
Rotor, cleaning	92
Runs	36, 186

S

Safety controls	45
Screwdrivers	11
Shack control	112
Short	21
circuits	21
cycling	100
Softwoods	70
Soot	51
Spread-type burners	134
Squirrel cage fan	82
Stack control safety switch	130
Stator, cleaning	92
State government energy incentives	325
Steam heating systems	31, 54
Stoves	206
buying	209, 212
circulating	214
coal	206
installation ideas	220
installing	212
maintenance	225
types	207-208
wood	206
Supplementary	
heat	254
heaters, types	174

T

Temperature controls	60
Thermocouples	134, 140
Thermostats	42, 180
anticipator	280
guidelines	290
location	282
maintenance	277
outdoor	256-259, 274-277
set-back	284
types	267-274
Tin snips	11, 188
Transformers	146, 152
Tuneups	104
furnace	104
oil furnaces	114
professional	123

U

Upflow horizontal furnaces	38

V

Voltage tester, making your own	14
Voltmeter	12, 28
dial meter	12-14
safety precautions	14
types	12

W

Wall heaters	53
Washable furnace filter	77
Wire cutters	11
Wood	31, 46
heat & efficiency	69
moisture content	70
stoves	206
Wood furnaces	46-48, 71, 226
duct systems	47
fan	47
Wrenches	11

Z

Zone temperature control	61